Innovation Never Stops

Also available from ASQ Quality Press:

The Executive Guide to Innovation: Turning Good Ideas into Great Results
Jane Keathley, Peter Merrill, Tracy Owens, Ian Meggarrey, and Kevin Posey

Continuous Permanent Improvement
Arun Hariharan

The Quality Toolbox, Second Edition
Nancy R. Tague

The Certified Manager of Quality/Organizational Excellence Handbook,
Fourth Edition
Russell T. Westcott, editor

The Executive Guide to Understanding and Implementing the Baldrige Criteria
Denis Leonard and Mac McGuire

Root Cause Analysis: Simplified Tools and Techniques, Second Edition
Bjørn Andersen and Tom Fagerhaug

The Executive Guide to Understanding and Implementing Quality Cost Programs
Douglas C. Wood

The ASQ Quality Improvement Pocket Guide: Basic History, Concepts, Tools, and Relationships
Grace L. Duffy, editor

The Quality Improvement Handbook, Second Edition
ASQ Quality Management Division, John E. Bauer, Grace L. Duffy, and Russell T. Westcott, editors

Making Change Work: Practical Tools for Overcoming Human Resistance to Change
Brien Palmer

Office Kaizen: Transforming Office Operations into a Strategic Competitive Advantage
William Lareau

Root Cause Analysis: The Core of Problem Solving and Corrective Action
Duke Okes

To request a complimentary catalog of ASQ Quality Press publications, call 800-248-1946, or visit our website at www.asq.org/quality-press.

Innovation Never Stops

Innovation Generation—the
Culture, Process, and Strategy

Peter Merrill

ASQ Quality Press
Milwaukee, Wisconsin

American Society for Quality, Quality Press, Milwaukee 53203
© 2015 by ASQ
All rights reserved. Published 2015
Printed in the United States of America
21 20 19 18 17 16 15   5 4 3 2 1

Library of Congress Cataloging-in-Publication Data

Merrill, Peter.
 [Innovation generation]
 Innovation never stops : innovation generation : the culture, process, and strategy / Peter Merrill.—Second edition.
    pages cm
 Earlier edition published as: Innovation generation : creating an innovation process and an innovative culture.
 Includes bibliographical references and index.
 ISBN 978-0-87389-912-3 (hard cover : alk. paper)
 1. New products. 2. Creative ability in business. 3. Technological innovations. I. Title.

HF5415.153.M47 2015
658.4'063—dc23
                                                                                        2015016050

ISBN 978-0-87389-912-3

No part of this book may be reproduced in any form or by any means, electronic, mechanical, photocopying, recording, or otherwise, without the prior written permission of the publisher.

Publisher: Lynelle Korte
Acquisitions Editor: Matt T. Meinholz
Managing Editor: Paul Daniel O'Mara
Production Administrator: Randall Benson

ASQ Mission: The American Society for Quality advances individual, organizational, and community excellence worldwide through learning, quality improvement, and knowledge exchange.

Attention Bookstores, Wholesalers, Schools, and Corporations: ASQ Quality Press books, video, audio, and software are available at quantity discounts with bulk purchases for business, educational, or instructional use. For information, please contact ASQ Quality Press at 800-248-1946, or write to ASQ Quality Press, P.O. Box 3005, Milwaukee, WI 53201-3005.

To place orders or to request ASQ membership information, call 800-248-1946. Visit our website at http://www.asq.org/quality-press.

 Printed on acid-free paper

Quality Press
600 N. Plankinton Ave.
Milwaukee, WI 53203-2914
E-mail: authors@asq.org

The Global Voice of Quality™

To my mother Phyllis Merrill, a remarkable lady, who taught me tenacity, the primary skill of the innovator.

To my daughters Rachel and Sarah, today's "innovation generation," who keep my ideas fresh.

To Angela, the other half of my life, whose creativity balances my focus.

To my late father William and grandfather Herbert, both of whom were writers, artists, and scientists.

Table of Contents

List of Figures and Tables . . . . *xiii*
Preface . . . . *xv*
Introduction . . . . *xix*

Part I: The Essentials

Chapter 1: The Why, What, and How 3
Creating the Future 4
Quality and Innovation 5
Strategy 7
The Mind of the Innovator 8
Convention and Innovation 8
The Process 9
Creativity and Innovation 11

Chapter 2: The Need for Knowledge 15
Leveraging Collective Knowledge 16
Individual and Organizational Learning 18
Information Flow 18
Network Theory 20
Knowledge Management 21
How We Grow Knowledge 22
    Stage 1 23
    Stage 2 23
    Stage 3 24
    Stage 4 24
Collective Knowledge 24
Taking a Bath 25

Chapter 3: The Roles and the Culture **27**
    Culture 28
    Knowing Your Best Contribution 29
    What Do the Scores Mean? 31
    Creator 32
    Connector 32
    Selecting the Solution 33
    Developers 34
    Doers 35
    Learn to Work with Others 35

Chapter 4: The Process and the System **37**
    The Opportunity 40
    The Solution 41
    The Tipping Point 41
    Development 42
    Delivery 42
    The Quality Management System and Innovation 43
    The Innovation Management System 45

Chapter 5: Innovation Strategy **47**
    Mega Trends 48
    Your Own Market Opportunity 49
    The Five Forces 50
    Start with Your Own Customers 52
    Deciding Where Not to Play 55

Part II: The Process

Chapter 6: Seeing the Opportunity **59**
    Staying Loose 60
    "The Scan" 62
    Big Data 63
    What Would Make Your Customer's Life Easier? 63
    The Pain Statement 64
    Start with Your Core Customers 65

Chapter 7: Connecting to the Solution **69**
    Defining the Problem 70
    Connecting 71
    The Myth of Epiphany 72
    Ideation 74
    The Sock Exercise 74

TRIZ	77
Sacred Attributes	78
Walk before You Run	79
Everyone Else Brings Data	80
Test Your Ideas	82

Chapter 8: The Tipping Point **83**

Selecting Solutions	84
Selection and Strategy	85
Short-Term Thinking	86
Beware of "Minor" Innovations	88
Internal and External Risk	91
Infrastructure	92
Risk Mitigation	92
Risk and Social Responsibility	93
Influence Chart	93
Switching from Loose to Tight	94

Chapter 9: Developing the Solution **97**

Development and Queuing Theory	99
The Development Steps	100
The Change from Loose to Tight	102
Behavior Change	104
Resistance to Change	105

Chapter 10: Delivery **109**

Value Proposition	112
The Statement of Value	113
Developing the Value Proposition	114
New Product Introduction	115
A Mind Map of the Innovation Process	117
Look for Other Market Opportunities	117
Your Innovation Process Capability	118

Part III: The Culture

Chapter 11: Culture and Behavior **123**

Behaviors Are Based on Values	125
Exploration	125
Interaction	127
Observation and Note Taking	127
Collaboration	128
Experimentation	128

Ability to Embrace Failure	129
Recognition of Behavior	130
Changing Behavior	130
Maintaining and Strengthening Values	131
Recruitment	131
Chapter 12: Innovation Teamwork	**133**
Communities of Innovation	134
Innovation Projects	135
Leading a Project Team	136
Early Results	137
Collaboration	138
Project Change: Loose to Tight	139
Team Culture Shift	139
Hanging Tight	140
Delivery	141
Chapter 13: The Competent Innovator	**143**
Leadership Workshop	143
Management Workshop	144
Innovation Competencies	146
Exploration and Interaction	146
Questioning and Listening	147
Note Taking and Technology	147
Networking	149
Ideation and Creative Problem Solving	149
Being Prepared to Experiment and to Fail	150
Give Yourself Thinking Space	151
The Gift of Time	151
Storytelling	152
Other Skills to Consider	153
"Bus Stop" Answer	154
Chapter 14: Networking	**155**
Networks of Friends	156
Successful People Will Network	157
Famous Collaborators	158
Different Types of Networks	158
The "Mirror" Trap	159
Your Personal Knowledge Network	162
Chapter 15: Leading Innovation	**165**
The Role of the Innovation Leader	165
Develop the Culture	167

Leading Creativity	168
Leading the Process and the Project	169
Leading the Process	169
Set the Challenge, Not the Vision	172
Resource the People	172
Entrepreneurship	173
Make the Decisions	174
Taking Action	175
Use a Project to Initiate Change	175

Part IV: The Implementation

Chapter 16: The Innovative Organization 181

A "Flexible" Organization	182
Growing the Organization	182
The "Spin-Off" Organization	184
New Knowledge, New Behavior	185
The Brand	185
KPIs in a Creative Organization	186
Structuring for Innovation	187
Using ISO 9001	187
Setting Strategy	188
Leadership, Risks, and Opportunities	189
Evaluating Risk	190
Where to Begin	191
Finding Solutions	192
Managing the System	192
Benefits, Not Features	193

Chapter 17: Measuring Innovation 195

Measuring Innovation Aptitude	196
Measuring the System	197
Measuring Processes	201
Measuring Product and Service	203
The Measurement Strategy	204

Chapter 18: Open Market Innovation 207

How Open Market Can Work	209
Stage 1: Creating the Opportunity	209
Stage 2: Finding the Solution	209
Stage 3: Making It User-Friendly	209
Stage 4: Execution	210
The Rewards	210

    Making Alliances Work 211
    Seek to Understand 211
    Network Alliances 212
    Real-Life Results 212

Chapter 19: Lighting the Fire 215
    The Change Team 216
    Changing to an Innovative Culture 216
    Communicate Vision 218
    Build Critical Mass 220
    Enable Action 221
    Raise the Bar 221
    Integrate the Win and the Change 221
    Recognize Success 222

Chapter 20: Creating the Future 223
    So, What Do We Do? 224
    Fulfillment for the Innovator 224
    Where Do I Begin? 225
    The Result 226
    What Do I Do Now? 226

Endnotes *229*
Index *233*

List of Figures and Tables

Figure 1.1	Primary process areas of business.	6
Figure 1.2	The innovation process—a paradox.	11
Figure 2.1	Traditional organizational structure.	18
Figure 2.2	Process map example.	18
Figure 2.3	Organizational networks.	20
Figure 2.4	Releasing knowledge.	22
Figure 2.5	The knowledge management cycle.	23
Figure 2.6	Transfer of tacit knowledge between people.	25
Figure 3.1	Creator, connector, developer, doer.	30
Figure 4.1	The stages of the innovation process.	39
Figure 4.2	The innovation cycle—the roles.	40
Figure 5.1	Competitive forces affecting survival of a business.	51
Figure 5.2	The hedgehog principle.	52
Figure 5.3	Launch, ramp up, exploit, reconfigure, disengage.	53
Figure 5.4	Author's business and personal "hassle" lists.	54
Figure 5.5	Reader's business and personal "hassle" lists.	54
Figure 6.1	Innovation process assessment.	66
Figure 7.1	Failure in classic problem solving.	73
Figure 7.2	Innovation process assessment.	81
Figure 8.1	Project risk versus time scale.	87
Table 8.1	Innovation focus by industry sector.	88
Figure 8.2	Project risk versus familiarity.	89
Figure 8.3	Assessment of risk.	90
Figure 8.4	Project risk versus ROI.	90
Figure 8.5	Influence chart.	94

Figure 8.6	Innovation process assessment.	95
Figure 9.1	Three-dimensional risk diagram.	103
Figure 9.2	Behavior change and innovation.	105
Figure 9.3	Innovation process assessment.	107
Figure 10.1	Pilot test procedure.	115
Figure 10.2	Innovation process assessment.	117
Figure 10.3	Mind map/strategy diagram.	119
Figure 12.1	The shift in team content and culture.	140
Figure 13.1	Innovation project tools and metrics workshop.	145
Figure 13.2	Mind map for an innovation process training session.	148
Figure 14.1	The spark of ingenuity.	159
Figure 14.2	Open and closed networks.	160
Figure 14.3	Hugo's circle of friends.	161
Figure 14.4	Carol's circle of friends.	161
Figure 14.5	Your own circle of friends.	162
Figure 15.1	More detailed process model—showing feedback	170
Figure 15.2	More detailed process model.	171
Figure 15.3	Leaders connectivity.	176
Figure 16.1	As the organization grows, it loses agility.	183
Figure 16.2	What's branding got to do with it?	186
Figure 16.3	Create "new knowledge" from your ISO 9001 management system.	191
Figure 17.1	Levels of measurement.	196
Figure 17.2	Organizational assessment.	198
Figure 17.3	Key issues.	201
Figure 17.4	Innovation process measurement plan.	202
Figure 17.5	Areas for measurement.	204
Figure 18.1	Sources of innovative ideas.	208
Figure 19.1	The road map to an innovative culture.	217
Figure 19.2	The change team plan.	219

Preface

The first edition of this book was written just six years ago, and it is remarkable how our world has changed in that short time. Stories I shared then seem dated now; companies that were at the pinnacle of success have entered grave difficulty. In the first edition of this book, I was explaining the "brave new world" of *innovation*, and believers were in a minority. After a few short years, the world of innovation is not just an R&D activity, it is everybody's job.

We have seen fundamental changes in the way we live as a result of innovation. Amazon majorly disrupted the retail world as they exploited the growth of online retailing to provide commodities that are nonperishable. Google's moves into phone operating systems upset the traditional phone businesses.

Big Data is everywhere. Vodafone, a UK telecom company, can pinpoint traffic jams based on how fast phone users are driving, and sells the information to TomTom, a GPS provider, to tell its own users where the jams are.

There is a fundamental change in higher education. It has become prohibitively expensive, and that is always a green light for the innovator. Online universities enroll, educate, and grant degrees to far more students at much lower cost than traditional universities.

Innovation is going through the "growing up" phases that quality went through 20 years ago, although, not surprisingly, it is growing up much faster. *Quality* left behind the myth that quality was the job of the quality department and became *quality management*. Innovation is leaving behind the myth that innovation is the job of R&D and is now discussed in terms of *innovation management*. A number of countries developed national guidance standards on innovation management, and these pointed to an International Standard on innovation management, ISO 50500, the responsibility of ISO Technical Committee TC 279. ASQ decided to create an Innovation

Division, and all these events beg the questions "Why innovation?" "Why now?"

In fact, we are in a "perfect storm" of change. The emerging nations of China and India are becoming major forces in the marketplace. Digitization means we do jobs now in an entirely different way than 10 years ago. Technology is enabling us to live longer, and we are dealing with the effects of an aging population and demands on healthcare. The downside of technology is that we have polluted our planet, and we are starting to see the effects of global warming.

Charles Darwin summed up the consequences of these changes when he said, "It is not the strongest of the species that will survive, nor the most intelligent, but those most able to respond to change." Our response to the complexity of change has emerged as *innovation*. The forces of change are recognized as the prime driver of innovation, and there is new material on this in the opening chapter and in Chapter 19.

The relationship of innovation and quality has been a controversial subject for some time, and this is also now explained in Chapter 1. Thinking on innovation has evolved to produce a remarkable consensus in a short time on the need for innovation management and a management system approach to innovation. This is also addressed in the opening chapter and more fully in Chapter 16.

Our understanding of creativity has grown, and this is discussed in Chapters 6 and 7, while Chapter 7 now has additional material on idea creation, or "ideation" as it is now called. There is new material on management of risk in Chapter 8, and this is tied to the metrics of innovation with new material in Chapter 17. Throughout the book there are many updates on a subject that has moved very quickly in a very short period of time. At the same time, there is certain "profound knowledge" described in the first edition and retained in this second edition.

We can all see how the events in our lives have drawn us to the place where we are at this moment. We have made many choices along the way. Some choices were tough, some were easy. The choices I have made have led me inextricably into innovation and have influenced why I wrote this book.

My grandfather and father were both artists, and they were both also writers. My grandfather was a chemist before he entered the church, and my father was an electrical engineer, and a member in the Institute of Electrical Engineers until he died. With those influences, I am a writer, an artist, and an engineer, and keynote speaker.

However, there is a creative component inside me that comes both from painting as a child and having been given a chemistry set at a young age. Chemistry took me down the road to becoming a chemical engineer, but in truth, what I enjoyed about chemistry was the incredibly creative and visual and sensory aspects of the chemistry lab. I guess that is where the desire to experiment began.

My exploring started in my teens. At the age of 12 I bought my cousin Bill's bicycle and started going places I had never gone before. At the age of 14 I cycled from the Midlands of England to the Lake District with my friends Dave, Bob, and Roger, and we were away for two weeks. I can't believe my parents allowed it to happen. The travel bug bit and has stayed with me my entire life.

Since I graduated from Birmingham University, the only part of my degree that still sticks is my thesis. It was on the use of "The Plasma Jet as a Chemical Reactor." I still have my original thesis after all these years, and interestingly, my primary finding as a raw undergraduate was that the science of plasma had been held back through a lack of collaboration between the players. A huge lesson for today's innovator.

From university I joined the Courtaulds Group. As well as chemical engineering, Courtaulds was a textile company. Textiles is an exciting world. I actually left Courtaulds for three years and ran my own fashion business. I rejoined Courtaulds and became chief executive of Christy, entering one of the most exciting periods of my life.

When I took responsibility for the Christy brand, I inherited 150 years of history. William Miller Christy invented the humble towel as we know it today. Although it is 150 years old, the Christy brand was an example of breakthrough innovation. Its invention followed the classic stages of innovation.

I learned these vital lessons: (1) even though the Christy brand was 150 years old, you can't stop innovating; 2) innovation does not have to be high in technology; 3) artistic and scientific innovation follow the same process. Christy was a great learning experience for me, and I will refer to it several times in this book. However, after three years I was presented with an opportunity that few would decline.

I left Courtaulds, and everyone was stunned. I was a company man. I went to work with Phil Crosby and entered another period of excitement and change. After I had been with Phil for close to five years, he retired, and I decided again to move on.

I set up my own consulting practice, which I have now had for over 20 years. During that time I have had the privilege of working with exciting companies. I have helped organizations develop their management systems by using the ISO 9001 framework, and I have at the same time become very involved in the ISO organization and am now the national chair for ISO TC 279 on innovation management.

Throughout this time I have also been closely involved in the American Society for Quality (ASQ). ASQ is unique in the world as a professional body, and as I write I am honored to be the first chair of the ASQ Innovation Division.

The other exciting aspect of running a consulting practice is the wide variety of organizations I have been able to work with. Financial institutions like AIG, design organizations like IBM and RIM (now called BlackBerry), manufacturing companies such as Solectron and Husky, and other companies in the pharmaceutical, food, retail, and distribution industries. As I've helped these people develop their management systems, the natural question once the system became effective was "Where to next?" I have been drawn into the need for these companies to be more innovative. I have keynoted on innovation at conferences, and trained and consulted with organizations seeking to develop their innovative ability.

When I first talked with Matt Meinholz at ASQ Quality Press about the title of the first edition of this book and proposed *Innovation Generation*, he asked, "Is that a noun or a verb?" My answer was, "It's both, actually." This takes you back to my first book *Do It Right the Second Time*, which focuses on the importance of balancing both people and process. We truly are an "innovation generation" (the noun), and "generating innovations" (the verb) is one of life's most rewarding experiences.

The world of innovation is exciting. Welcome to the future, and be aware of the big lesson of this book, which is now its main title: *innovation never stops*.

Peter Merrill
Kilbride, Ontario
Canada
January 1, 2015

Introduction

A BRIEF HISTORY OF INNOVATION

If you look at the "spikes" in knowledge growth throughout the recorded history of the Mediterranean, Europe, and North America, these spikes are getting closer together. Ancient Egypt spiked in 3000 BC, and the effects of that knowledge explosion lasted for more than half of our recorded history, until 1000 BC. The Greco-Roman knowledge explosion spiked with Alexander and Aristotle in 400 BC. This lasted less than 1000 years, until the fall of Rome in the fourth century AD.

We emerged from the dark ages with the Renaissance and the work of giants such as Leonardo da Vinci and Michelangelo. A mere 300 years later came the Industrial Revolution, and the next big spike in knowledge growth was between 1880 and 1920 with the era of the "inventor." As the United States emerged from the death and carnage of the Civil War and people refocused on living, communities across the country started to grow, and the industrial revolution gathered pace along with business and trade.

People needed to travel to other communities to trade, or better still, talk to people in other communities without the traveling. They needed to spend time in the evening with their families as their work increasingly took them away from home.

Henry Ford saw that the need for speedy travel would not be solved with faster horses, Alexander Graham Bell saw that the need to communicate was not being solved with the telegraph, and Thomas Edison saw the shortcomings of oil and gas lighting in the home.

Just 100 years later we are in the "innovation generation," and another explosion of knowledge through information technology. We are in an age where people are expecting us to innovate. It's not just nice to come up with new ideas—customers and consumers are expecting it, and are actively "chasing cool."

Unfortunately, business has not positioned itself well for innovation. In the last years of the 20th century, business focused on improving the delivery of existing products and services and less on the development of new markets and new products. As a result, quality management has frequently become internally focused. The harsh reality is that if a product or service has become outdated, effort to improve its delivery efficiency is totally wasted.

Innovation is about developing the products and services that the market needs tomorrow, and is driven by the need for convenience, not by technology. Finding a "cool new idea," while it may be interesting and exciting, has no value unless the idea solves a real problem. Even then, new ideas will only be adopted if they are easy to adopt.

Contrary to popular belief, new ideas emerge most frequently as the result of collective knowledge, and typically do not come from a lone "genius." Consequently, for successful innovation to occur, the internal and external networks of an organization have to be well developed. An innovative organization enables information to flow freely between its people and also enables information to flow freely to and from external organizations. This type of "networked" organization transmits information and knowledge rapidly and effectively.

Interestingly, documented knowledge typically accounts for only 20 percent of the knowledge in an organization. Good *knowledge management* (KM) is the platform from which innovation is developed, and an innovative organization enables the rapid transfer of the knowledge that is in people's minds by using its network. To innovate, we have to release the knowledge in people's minds, and "a culture of innovation" is essential for doing this.

We all have a role in the innovation process. *Creators* generate opportunities, *connectors* link opportunities to solutions, *developers* make solutions practical, and *doers* deliver solutions. Through the self-assessment in this book, you and your colleagues can identify where you make your best contributions to innovation.

Start your innovation process with core customers, then move on to "not yet" customers. Engaging people in a good innovation strategy will find new markets where there is no competition. However, you must position yourself quickly so that the competition later sees no value in entering your market.

THE PROCESS

Many people are surprised that innovation follows a defined process, and that creativity is only part of the innovation process.

The process is driven by need, and not by technology. Surprisingly, the market may not realize that the need exists. You need to ask customers questions like "What are your hassles?" and "Where do you waste time?" and also remember that it is easier to change your product than to change your customer base. This "creation" work takes time, and breakthroughs usually occur after a lot of hard work, even though the mode of operation is loose at this early stage. Brainstorming with customers is one of the primary techniques for finding these opportunities.

When we move to the solution stage of the process, we find that "connecting" to a different environment frequently reveals a "radical" solution, and we have learned, as Linus Pauling said, "The best way to have a good idea is to have a lot of ideas . . . ." Using collective knowledge is the most powerful way of finding alternative solutions.

Most of the solutions you find are built on a lot of previous experience, and an "epiphany" is really the last piece of the jigsaw puzzle. However, there is one other "pain point" you must go through. Somebody will have to say "You can't do that" when you offer a radical solution.

The tipping point in the innovation process is when we select which options to pursue, but, sadly, choices are often made based on poor data, and organizations frequently try to pursue too many options. Organizations are also drawn too often into safe and minor innovations; they have become "risk averse," and only select short-term choices.

A new-product portfolio must include a number of long-term and potentially major innovations in order for an organization to have a healthy future. Risk-taking is fundamental to innovation.

The selection stage or tipping point is also the point where the process switches from "loose" to the "tight" mode required for the development stage, and organizations often have difficulty with this essential change. The primary challenge at the development stage is to make the new product or service easy to use. To most people's surprise, you will generally need an exponential improvement in time or cost savings for rapid acceptance of a new product. Partnerships with the supply chain and delivery chain are also vital here. At this development stage, time is of the essence. The chance is that someone else on the planet has already found a solution that will compete with your own!

Executing the final working solution is the toughest challenge of all, and research by the Conference Board has shown that of 3000 original ideas, only one survives. Engaging operations and sales people in the development stage prepares them for this delivery stage. The best ideas don't always make it, and competitors with inferior offerings will copy you if you are not fast to market. The "value proposition" has to link to the customer's previous experience and show just two or three key advantages for your new product. This way, the customer will "get it."

THE CULTURE

I'm frequently asked, "How do we develop an innovation culture?" The reason this is a challenge is that much of what we have implemented in our organization over the last 20 years has inhibited the creative aspect of innovation. The second reason for difficulty is that there are in fact two cultures.

Culture is based on behavior, and the creative phase of innovation requires the behaviors of exploration, collaboration, and experimentation. We explore less as we have become attached to our desk and computer screen. Although we think otherwise, we collaborate less because the bandwidth of e-communication is far less than the bandwidth of interpersonal communication. We experiment less as we have become increasingly risk averse. Clearly, these behaviors have to be restored, and I will show you how.

The coexistence of creative behavior with an "execution" culture requires careful managing. It is too easy for the execution culture to kill the creative culture. Separate organizations are not the answer, and I will show you how to transition between the two sets of behavior as you move through an innovation project.

Supporting these behaviors of the "community of innovation" is something for an organization to consider in the early days of culture development. This community grows the knowledge and understanding needed for both the organization and for individual projects. ASQ did this with its Innovation Interest Group, which grew rapidly into the Innovation Division.

INNOVATION NEVER STOPS

A leader's most immediate task is to set a direction for innovation, and a leader has to make tough decisions on which ideas to pursue. Direction is derived from market need and the core competencies of the organization. Rest assured that direction will change. To manage changes, "change agents" need to plan for change, and you start the change by focusing on a product, service, or market segment with declining revenue.

The organizational assessment in this book will enable you to identify the areas of weakness in your organization and develop an innovation plan. As General Eisenhower remarked, ". . . I heard long ago in the Army: plans are worthless, but planning is everything." The early win is vital in order to light the fire of innovation and join the innovation generation. However, your plan will change.

As the innovation process starts working, the organization should understand that it is a cycle. The lifetime of knowledge shrinks in periods of rapid innovation, and as the pace of change increases, we are becoming increasingly aware of the "need for speed," and that innovation never stops.

Part I

The Essentials

1
The Why, What, and How

The future is very important; it is where we will spend the rest of our lives.

—Joel Arthur Barker

It was midsummer, and I was visiting my family in England. I had just left the Midlands and I was driving on the M40 motorway between Oxford and London. It was a beautiful clear day, and then in the corner of my eye I saw a small red light start flashing. I glanced down at my BlackBerry device, a gift from Gary Cort, VP Quality at Research In Motion (RIM), and my good friends at RIM. I had a message.

I quickly read the message on the screen. It was quite simple. My friend Armando Espinosa had e-mailed me from Mexico City and was asking if I could speak at an upcoming conference in Mexico. He wanted me to speak on "Quality Management and Innovation." Doing some mental arithmetic, I worked out that the conference would be on Labor Day weekend, and later confirmed with Armando that I would be delighted to accept his invitation.

As I drove the 10 miles to the exit I thought to myself how this remarkable little event was a testament to modern ingenuity and innovation.

A guy in Mexico City can send a message to a guy who he thinks is in Canada, but is actually in England, and minutes later will get a positive response. That simple little device, the BlackBerry, had enabled all of this to happen. This "world shrinkage" is a key factor in the world of innovation.

However, that story was a little over seven years ago. In that short time BlackBerry went into major tailspin, and advances in software now enable me to listen to my smartphone message instead of reading it, and then speak my response, translated through software into text. We are in a world of innovation, and its speed increases every day. Kodak invented the digital camera back in the '70s and ignored it. They no longer exist. Nokia invented the tablet computer in 2004 and ignored it. They also went into decline.

We have seen fundamental changes in the way we live as a result of innovation. Amazon has majorly disrupted the retail world as they exploit the growth of online retailing to provide nonperishable commodities. It absorbed companies like Zappos the shoe people and Qudsi the diaper people. Amazon then set the challenge of delivering its products more rapidly, and hence it's pursuit of the use of drones. Innovation never stops.

There has been a fundamental change in higher education. It has become prohibitively expensive, and that is always a green light for the innovator. Online universities enroll, educate, and grant degrees to far more students at much lower cost than traditional universities. E-learning reaches far more people than in the largest lecture hall. Initially, e-learning was not good, but it has improved while holding cost and convenience.

Shai Reshef, who built his University of the People to offer affordable online qualifications to anyone, describes it in this way, "I realized that everything that made higher education so expensive was already available and for free." The assumption that the components of successful innovation are "already available" should be liberating for companies. At the same time, recognize that, like many innovations, the fundamental concept is not new. The Open University was created in the UK by Prime Minister Harold Wilson in 1969. He regarded it as his greatest achievement in making higher education open to all. You will see their distinctive logo of a big "O" on a shield on educational TV programs.

Students do value the social aspects of a "live" college—the freedom of living away from home and the friendships—that e-learning can not provide. As a result, e-learning offers a blend of online and on-campus courses, but this forces them toward the cost of traditional universities. Eventually, they may solve this problem, but at present they fall short of satisfying all students. You can never stop innovating

Innovation is exciting. It is a new world. Innovation is about developing the products and services that the market needs tomorrow. It is about creating the future. As Joel Arthur Barker said, "The future is very important; it is where we will spend the rest of our lives." However, a word of warning. Innovation never stops, and the company that stops innovating will die.

CREATING THE FUTURE

Disruptive innovation shows itself throughout history, with the invention of the steam engine in the eighteenth century, the use of electricity in the nineteenth century, and the digital revolution at the end of the twentieth century. These innovations not only changed the way things got done, but also opened the door to more change and more opportunity for innovation. This is that perfect storm we find ourselves in as we enter the twenty-first century.

Change is coming at us from more directions than ever before. The marketplace is changing as China and India, with 35% of the World's population, become global players. Our population is itself changing as technology enables us to live longer, and technology is changing at a pace faster than ever before. Not all the change in technology is beneficial, as we recognize the planet is warming, for example.

In the Preface I mentioned Charles Darwin, who spoke to this problem 150 years ago when he said, "It is not the strongest of the species that will survive, nor the most intelligent, but the one most responsive to change." It is the ability to respond to change that is today's challenge. The innovator responds to change.

Recognizing the major trends creates the best opportunities for innovation, whether these trends are globalization, information technology, global warming, or an aging population. New trends create new needs. New needs create new problems, and so create new opportunities for the innovator. The innovator is less interested in the effect of these trends on their own organization and more interested in the effect of these trends on the customer. Change for the customer creates new opportunity, and within your own world there are many opportunities for innovation.

Innovation is about developing the right things, the products and services that the marketplace truly will need tomorrow.

QUALITY AND INNOVATION

Organizations exist because they enable someone else (the customer) to get something done. Just *satisfying* the customer leaves you open to the competition. *Delighting* the customer means they are happy and unlikely to move to the competition. The easier they can do their job, the more delighted they will be.

The true quality professional recognizes that there are always customer needs that have not been met. In order to "fulfill" or "completely fill" the needs of the customer, we must seek to deliver a solution that is probably beyond the existing capabilities of our organization. The quality professional who has vision wants to address these unfulfilled needs and does this by innovating. I invite you to broaden your vision beyond quality management to innovation management.

We find innovation opportunities when we go back to the customer to find out which needs have not been fulfilled. These unfulfilled needs usually occur because we have not adequately defined the problem.

For business success in that future world, we must be thinking about tomorrow's customer and what their needs will be. Much of business has tended to focus on today's customer and address their immediate concerns. Business has also focused inwardly, and this has caused us to neglect the future. In the last half of the twentieth century, business focused heavily on improving delivery of its products and services. This focus was supported by investment in just two of the four main business process areas: *infrastructure*, through investment in information technology, and *product delivery*, through investment in quality management.

Referring to Figure 1.1, the four primary process areas of a business are (1) market development, (2) product development, and (3) product delivery, which are held together by (4) infrastructure.

You can see that for a balanced business that is expecting to have long-term success, the investment in delivery and infrastructure needs to be balanced with investment in market development and product development. That means investment in innovation.

Quality management talks about customer focus, but much of this focus has been internal process focus and needs to move back to a market focus. In the world of quality management, people have been

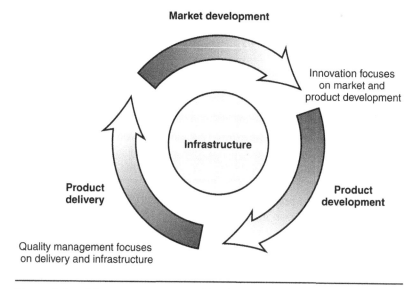

Figure 1.1   Primary process areas of business.

schooled over the last 30 years to "do it right the first time." However, if you are not *doing the right things* and your product or service has become out of date or "mature," then your effort to improve efficiency is wasted.

Innovation is about developing the *right things*, the products and services that the market needs tomorrow.

STRATEGY

A good innovation strategy will put you ahead of the competition, in a future position where the competition sees no value in entering your market. Cirque du Soleil created a unique market niche by innovating the tired concept of the circus, and companies chasing Apple arrive at the market space and find that Apple has moved on. Back in the '80s, Milliken was the benchmark in quality management, and it was their embrace of the "quality religion" that drew my own corporation, Courtaulds, also a chemical and textile company, into quality management. Milliken is now a world leader in advanced materials, and left textiles in the '90s. They never stopped innovating. These are outstanding innovators. This book will explain how these and other similar companies executed their strategy.

Innovation is a critical component in any overall business strategy, and it was Michael Porter who brought strategy into fashion.[1] His "five forces" model was simple. His critical insight is "corporate strategy should meet the opportunities and threats that exist in the external environment." Porter identified five competitive forces that shape every industry and every market. They are the threats from *suppliers, customers, new entrants, substitutes*, and *rivals*.

If we look at Porter's work on strategy, his competitive forces *suppliers* and *customers* actually become the allies of the innovator. I will explain this in Chapter 6, "Creating the Opportunity" and then later in Chapter 18, "Open Market Innovation." In periods of rapid knowledge growth such as we are experiencing today, there is more knowledge "outside the box." You will be able to access this knowledge most easily by sharing your own knowledge with your business partners. Even Porter's work has been superseded, as Rita McGrath shows us that competitive advantage is only transient given the speed of change.[2]

If you think competitive threats are only on the home front, consider the "hidden dragon." Many of today's products, especially software, were developed in Asia. If you think low cost is the threat, then think again. It is low income, not low cost, in China and India that drives innovation. India developed the world's largest mobile phone market at 20 percent of Western cost, and China is responsible for more than half of global production of motorcycles. These results were achieved through innovative approaches to service delivery and innovative process networks. North America needs to ask itself, "What is next after the Japanese automobile threat? " Once more, the customer drives innovation, not technology. Technology is just one way of getting to a solution.

Innovation is about addressing all of these threats.

THE MIND OF THE INNOVATOR

The popular perception of the mind of the innovator is that of the person who finds a cool new idea. Unfortunately there have been many so-called innovative answers or solutions to problems that turned out to be of little significance in the marketplace.

People ponder on a problem that impacts them personally and go through a "connecting" experience. They have that "Aha!" moment or epiphany. The excitement of the epiphany leads them to believe that the problem is more important than they had thought. They have been thinking about the problem for a long time, so it has increased its importance in their mind. In truth, they find the answer to a problem, and yet the more important question is "How important is the problem?" Too many large organizations have departments creating solutions looking for a problem.

At the other end of the process, having found a potential solution, people all too often forget about the user of the solution. They rush to market or initiate a new service or product and upset the customer with an offering that is full of glitches and bugs. It's like the old days of buying a new car, or today buying new software.

Making an innovation user-friendly is a vital step in the process. If you don't, you will upset the market, and you will rarely get a second chance. More likely, a competitor will see your failure, address your shortcomings, and take your opportunity away from you.

CONVENTION AND INNOVATION

Business wisdom says listen to the customer, that technology drives change, and that you should recognize success. Convention also says target larger markets, seek higher margins, and avoid failure.

These principles all change for the successful innovator. Your market doesn't yet exist, so you must anticipate the market ahead of the customer. You must understand that market convenience, not technology, drives change. Additionally, you must plan for failure and learn from it.

We are surrounded by the results of convenience driving innovation. Photographic film has all but given way to digital photography. Notebook computers are used far less while traveling with the spread of smartphones and tablets. Greeting card shops are under threat as free Internet cards become available. In future years, electric companies may be downsizing as our solar cells and fuel cells replace them.

I recently had arthroscopic surgery on my knee, the consequence of playing football for too many years. I was able to go into hospital for day surgery and walk to my car afterward. In the past, this surgery would have required a week in hospital, a month on crutches, and several months of low mobility. I like this kind of convenience.

There are many myths surrounding the world of innovation.

It is often thought that innovation comes from a sudden stunning insight, but if you look at history, most innovations build on a lot of previous experience. People also believe that ideas come from a lone genius, whereas in fact most innovations are the result of collective knowledge.

People think that good ideas are hard to find and that they arrive randomly. In fact, there is no shortage of ideas. The bigger challenge is identifying the right opportunities.

The final reality of innovation is that commercialization is a major challenge, and most new ideas struggle to see daylight because potential users resist anything new. In truth, the best ideas don't always make it. I will keep emphasizing the "need for speed." You will need to have developed an effective quality management system with lean processes if you want to deliver your new product on time and to customer requirements.

Some people also believe that there is no innovation process, whereas the reality is that many innovations actually fight their way into existence through a process that is so poorly defined people don't realize it exists.

THE PROCESS

Phil Crosby used to say, "All work is a process," and yes, innovation is a process.

Innovation starts with identifying an unmet customer or market need. Importantly for the innovator, neither the customer nor the market may recognize that need. Henry Ford said, "If I had asked people what they wanted, they would have said 'faster horses.'" Creative thinking is usually required at this first stage. I have had a fortune cookie note on my notice board for years that says "The secret of a good opportunity is recognizing it."

Once the customer opportunity has been found, the problem solving is where most people recognize innovation as taking place. Breakthrough innovation comes from finding radical solutions. This is where connecting with a product or process from a totally different environment often leads to that "Aha!" moment. This is exciting! This is the creation of new knowledge. Henry Ford's ideas for mass production came from seeing a meat processing factory and applying the factory's concepts to production of motor vehicles.

However, after this second stage we only have a *concept*. We now have to develop a working solution. This is where the developers take over. The third stage in the process of making the solution work is where so many organizations lose momentum and lose the advantage they gained in stages one and two. Speed to market is essential. Discipline becomes vital. This is where Thomas A. Edison's saying "Genius is one percent inspiration and ninety-nine percent perspiration" becomes reality.

Finally, getting to market is where far too many stumble. The production, service delivery, and sales people need to have been involved in the earlier stages if you want a long-term and continuous innovation process. They now have to "run for the line." Every advantage we can give them is essential. Time and advance notice are their biggest advantages of all. Too many companies wrap their development activity in excessive secrecy, which slows down the process, and the sales and operations people are then presented with a challenge of which they had no previous knowledge.

The activities that need to be performed in an innovation process are:

1. Find the market need and opportunity
2. Identify potential solutions for the need
3. Select the solution
4. Develop the solution to be user-friendly
5. Take the solution to market

The organizational functions that perform these activities typically are:

1. Market research
2. Product and service research
3. Strategy
4. Product and service development
5. Operations and sales

In most businesses the second function, which I referred to as product and service research, is often the weakest.

In truth, quality professionals, with their inherent problem-solving skills, fit well in this area. However, they are limited by "left brain" analytical problem solving. In Chapter 7 I will introduce you to "right brain" creative problem solving, often called "ideation." This is how the innovator finds radical solutions or ideas. One other point to note: our focus on process management has led to linear thinking and has inhibited our ability to ideate.

To innovate successfully, an organization needs two modes of behavior in its innovation process: "staying loose" to create and conceptualize, and "hanging tight" to develop the concept and get to market (see Figure 1.2).

I will explain how to overcome this apparent paradox in the subsequent chapters.

CREATIVITY AND INNOVATION

People frequently confuse *creativity* and *innovation*. Creativity is only one of the attributes required for successful innovation. Creativity is a way of accessing new knowledge. It is also a way of retrieving existing knowledge that we may have forgotten, which we call *subconscious knowledge*. Creativity and innovation are different. Creativity is what initiates the innovation process, and people frequently think this is the whole process. I was a member of the Terms and Definitions working group that wrote ISO 9000. That experience developed my ability with terminology.

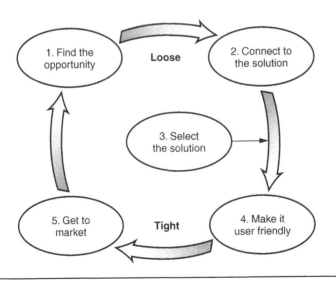

Figure 1.2   The innovation process—a paradox.

No definition is absolute, and most definitions in the English language have their roots in Latin, French, German, or Danish. When we look at the word *innovation*, the word *nova*, or *new*, leaps out at us. On the other hand, *creativity* is clearly about *creation*; this is about *building* or *developing* and is less about the final result.

Creativity is the ability to produce new ideas through imagination and through unconventional approaches to problems. Creativity occurs in an environment where people have freedom to think and interact with new stimuli. Sadly, the opportunity to create is removed from most of us as a result of the daily work we do, and yet creativity is the vital initiator of the innovation process.

You have seen that creating the opportunity and the potential solution are only the early stages of the process. Making the solution user-friendly and getting it to market are just as important stages in the innovation process.

As a result of what I have explained, we have a variety of definitions for innovation. The word "new" is fundamental. What becomes evident is that the fuel of the innovation process is "new knowledge." The final outcome can be a product or service, or it can be more esoteric and just be called "value." However, the word *value*, due to its financial implications, is giving way to "benefit." ISO/TC 279 has replaced *value* with *benefit*. Innovation is not just about money. Here are some key points:

a. *Renewal* is the etymological origin of innovation

b. *Novelty* and *introduction* are the two basic concepts

c. Economic benefits and societal betterment have emerged as key characteristics

d. Innovation is both an outcome and an activity

Also be aware that many "definitions" are in fact descriptions. A good definition follows the substitution rule whereby the definition can be inserted into a text to replace a term. A good definition will be about 10 words.

This gives us some potential definitions for innovation and for the innovation process:

- Conversion of new knowledge into new products and services
- Conversion of new knowledge into new benefit
- Creation and introduction of a new offering with benefit
- Something new or changed, realizing *benefits*

However, an emerging understanding of innovation is that true innovation will cause the user of the innovation to "do things differently." It is not the product but the effect of the product or service that matters. Innovation is different from improvement in that it causes a different behavior in the user, although at the same time, the user must not experience difficulty adopting the new behavior.

Innovation involves the successful implementation of creative ideas. Creativity is about *coming up with ideas*, while innovation is about both *creating ideas* and *putting those ideas into practice*.

This means that an innovation can be a change in your business model or business processes, or it can be a change in what you deliver to the customer, whether product or service.

The other challenge is to define whether innovations are "breakthrough," "disruptive," or "radical." Given the change in behavior that is implied, radical innovation can be described as a new offering that causes an activity to be carried out in an entirely different manner. Research has suggested that only about 15%, or one in six, of so-called innovations are radical. However, it is important to remember that the new behavior must be easily adopted. If you have to "train" your customer to use your new offering, your innovation is incomplete.

A final thought: Morgan Stanley found that companies with "radical" innovations generate 10 times more shareholder value in an IPO.[3] That emphasizes the need for creativity. The execution phase needs speed. An Accenture survey found that those who commercialize patents quickly outperform the market by 1000% over 10 years.[4] This emphasizes the need for speed to market.

In Chapter 2 we'll look a little deeper at this thing called "knowledge" that is so fundamental for innovation. Let me now draw your attention to the key points from Chapter 1.

KEY POINTS

- In recent years business has focused on improving the delivery of today's products and services and not on the development of new markets and products.

- Innovation is about developing the products and services that the market needs tomorrow.

Continued

Continued

- Quality management has frequently become internally focused, leading to "linear thinking" and inhibited innovation.
- If your product or service has become outdated, effort to improve efficiency is wasted.
- Good innovation strategy finds a new market, which the competition later sees no value in entering.
- Innovators must recognize mega trends such as globalization, information technology, global warming, and aging populations.
- Finding a "cool new idea" has no value unless the problem that the idea solves has significance.
- New ideas will not be adopted unless they are easy to adopt.
- New ideas are most frequently the result of collective knowledge and do not come from a "lone genius."
- Innovation is driven by the need of convenience.
- Innovation follows a defined process.
- Creativity is only part of the innovation process.

2

The Need for Knowledge

The only competitive edge an organization has is the ability to learn faster than the competition.

—Arie de Guess in Peter Senge's
book *The Fifth Discipline*

The twin towns of Kitchener and Waterloo celebrate the month of October with a Bavarian Oktoberfest, and October in Ontario is a beautiful time of the year. The leaves are turning and the colors are magnificent. The Kitchener-Waterloo section of ASQ invited me to speak at their October meeting and give my keynote speech "You Too Can Innovate," which I had given at the ASQ World Conference. The early evening drive from my home in Kilbride was an absolute pleasure. The colors of the trees brought out the artist in me.

The evening was a pleasure, and it was good to reconnect with old friends. The audience enjoyed doing the innovator self-assessment, which is part of the speech (and is described in Chapter 3). At the end of the evening the section gave me a gift of the book *Leonardo da Vinci* by Frank Zollner.[1]

Reading the book the following weekend reminded me how at the time of the Renaissance we had not separated art and science. Leonardo was a writer, an artist, and an engineer. He is someone with whom I have a great affinity. As I read the book I marveled at his understanding of the laws of physics and at the same time his magical ability to work with color.

If you saw the movie *The Da Vinci Code*, you will be familiar with the balance that he brings to his painting of *The Last Supper*. Most of us also marvel at the subtle smile of the Mona Lisa, which is captured in her lips and eyes. In another of his works, *Madonna and Child*, the colors are also subtle but very well balanced. The words *subtle* and *balance* keep appearing.

Many people think the artist and the engineer are poles apart. One is loose and creative, the other is tight and analytical. The engineer takes an activity that was previously shrouded in mystery and through their analytical and mathematical skills breaks it down into logical steps. On the other hand, the engineer is frequently critical of how the artist creates ideas out of "thin air" because these ideas are often intuitive and therefore do not appear to be substantiated.

Over the last century our educational system has increasingly separated these two views of life. It is convenient for the educator. It is not good for society. The fracture this has caused is also at the root of many of today's organizational roadblocks to innovation. Creative and analytical people are separated during their education, and frequently remain separated during their subsequent work life. For successful innovation, we have to reconnect the minds of the creator and the analyst. The left and right brains have equally important roles in the innovation process. The work of Leonardo da Vinci shows the importance of this balance. The challenge for the innovator is to create new knowledge, and this requires the left and right brains of both the individual and the organization to be truly connected.

LEVERAGING COLLECTIVE KNOWLEDGE

Information and knowledge must flow throughout an organization before the organization can become innovative. For information and knowledge to flow, the people in the organization must be linked.

A few years ago my mother experienced a fall and cut her leg. The hospital where she was treated, ironically named Good Hope Hospital, exacerbated her illness because the flow of information between people was so poor as to make the hospital organizationally incompetent. The individual staff and physicians were technically outstanding, but the hospital did not operate as a system. A primary problem was the inability of the organization to interface with external organizations. My mother contracted MRSA, which results from poor cleaning services. An organization is ultimately responsible for the results of all its subcontractors. This organizational incompetence showed up again in the inability of the hospital to interface with social services. A hospital that was supposedly short of beds took a full week to discharge my mother after she was ready to return home. If you are not organizationally competent, you can't be organizationally innovative. You may have many separate individuals "innovating," but until they come together collectively, you will not beat the competition.

The challenge of running a large organization is made more difficult by the fact that we are also dealing with complexity. We are starting to understand complexity in its many guises, whether we refer to systems thinking, Euler's network theory, or Gleick's chaos theory. We understand that a system operates as a result of the links between its elements. Gleick's book *Chaos: Making a New Science* brought this down to earth with the famous term the *butterfly effect*: "If a butterfly flaps its wings in Singapore it can cause a hurricane in the Caribbean." What this means is that cause and effect are no longer immediately related in a complex environment.[2]

The most powerful solution to this problem has emerged through the use of collective knowledge to develop our creativity and in turn produce innovative solutions to the problems created by change in complex environments.

Our understanding of how organizations operate has increased as we have understood better the principles of network theory and systems thinking. We now understand that for innovation to be effective, we need an *innovation management system*. We are starting to see these develop, and I will explain them more in Part IV.

An effective management system enables organizational innovation, and this is achieved through the flow of knowledge *between* people. To achieve rapid flow of information and knowledge and become a competent organization, the organization's processes, its people, and its technology need to come together as a system. This brings together the "left and right brain" of the organization.

Too often, businesses try to cover their management system deficiencies by trying to dazzle the customer with a wonderful new product or service in the hope that they will be forgiven their previous sins. However, this type of innovation is only "skin deep." This is a short-term strategy, and their customers will find them out when they fail to deliver. They may be creative, but they are not innovative.

So, the question is "How do we develop information flow in order to create organizational knowledge and apply this knowledge to become an innovative organization?" I am going to develop your understanding of how to create new knowledge, and also how to manage new knowledge. I am going to do this at both an individual and an organizational level.

INDIVIDUAL AND ORGANIZATIONAL LEARNING

At an individual level, we gain information and we process this information to create knowledge. We call this process *learning*. Many individuals do not progress beyond learning. They merely become "sources of knowledge."

Successful individuals, however, *apply* their knowledge in a way that creates value for themselves and others. Their knowledge flows out from them in a way that "makes a difference." They are competent and are strongly positioned to innovate. Competence is defined as *the ability to apply knowledge to achieve results*.

You can read as many books on golf as you wish, but until you have been on a golf course, you can't play golf. Talk to any golfer, and they will also tell you that you can't be competent until you have played in a competition. Once you are a competent golfer, you can innovate and try new ideas that will improve your game. Individual and organizational innovation are similar. Organizations must also be competent in order to be competitive. To achieve the necessary flow of information and knowledge to become a learning organization, the organization's processes need to come together as a management system.

When information flows between processes, organizational innovation is achieved through the application of organizational knowledge. To quote a good, and sadly departed, friend of mine, Herve Mignot from Paris, "The organization's head is connected to its feet."[3]

INFORMATION FLOW

For information to flow quickly, we need the right organizational structure. The traditional organization structure inhibits information flow (see Figure 2.1).

In the past we have attempted to address the problem of information flow by viewing activities from a process approach and by process mapping. I think most implementers of quality management have process mapped at some time. Figure 2.2 shows a typical process map example.

Process mapping is a technique adopted from the chemical industry. It has a lot of benefits, but it treats your organization as if it were a chemical plant, with processes represented by tanks connected by pipes. It only addresses the process aspects of your organization. If you have process mapped using the "swim lane" technique, which identifies people's responsibilities, then this gives you not only your process network but also the beginning of your "people network."

Chapter 2: The Need for Knowledge

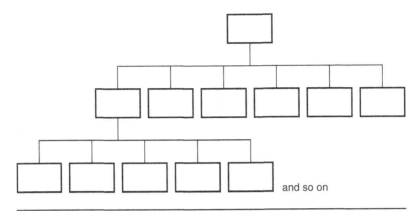

Figure 2.1 Traditional organizational structure.

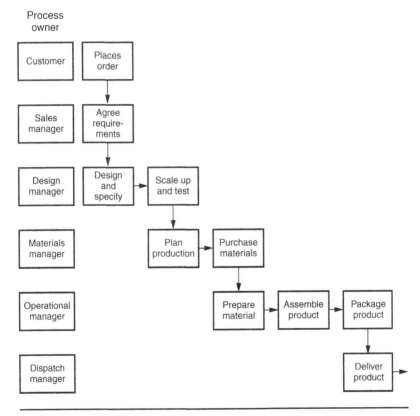

Figure 2.2 Process map example.

We have moved to systems thinking as we have process mapped our product flow. However, we have rarely mapped the flow of information between people.

NETWORK THEORY

This information flow between people is often referred to as the "network" of the organization, and in the past this network has relied far too much on random communication. An organization is a complex system. *Network theory* addresses the flow of information in complex systems. If we learn about the information flow in our network, it can enable us to manage our organizational knowledge.

Recent years have seen a tremendous growth in the understanding of network theory. You know that every organization has its own people network and how information flows between them. You need to understand the structure of your organization from a network perspective (see Figure 2.3).

This line of thinking was developed by Albert-László Barabási in his excellent book *Linked*.[4] I picked this book up in, of all places, the gallery bookstore after visiting my client Deborah O'Leary at the Art Gallery of Ontario. Barabási develops network theory by comparing social, technical, and biological networks. He shows how we originally thought all networks

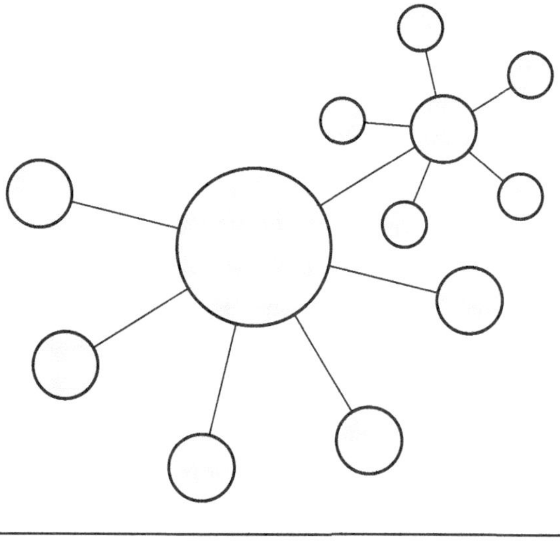

Figure 2.3 Organizational networks.

were random, rather like a highway system with a small number of links to each of its nodes.

The key points in a network are its *hubs* and *nodes*. In the people network of an organization, these are its managers, its meetings, and social activities, whether formal or informal. In a successful organization formal and informal networks are the same; it is an organism, like a human body.

Think of an air traffic network. Nodes (local airports) have a few links whereas hubs are highly connected. Take out a hub (major airport) and the network can be seriously damaged. Think about your own organization and how this applies.

The reason organizations experience so much difficulty with communication is because they do not understand their network structure. This structure is not the organizational chart, and it is not even the process map. It is the links between the people, or what we refer to casually as our *network*. You need to learn the primary information flows between the people in your organization. You can then understand how people in your organization gain knowledge, and you can then manage your knowledge. Techniques such as *social network analysis* enable you to do this.

Once you understand your information flow, you can start to grow your organizational knowledge, and that knowledge creates new products and services.

KNOWLEDGE MANAGEMENT

The innovation process is about the conversion of new knowledge into new products and services, so an innovative culture has to be a learning culture. We are acquiring new knowledge all the time, and we store this knowledge in our minds. An innovative organization must be agile, so we must have the courage to allow knowledge to be stored in the minds of our people and must not be obsessive about documenting knowledge.

The advances in information technology make this documenting very tempting. However, documentation seriously reduces the agility of an organization, and we can only document a fraction of the knowledge in people's minds.

Documented knowledge is referred to as *explicit* knowledge and comprises less than 20 percent of our available knowledge. The knowledge in our minds is *tacit* knowledge. There is a whole raft of untapped knowledge stored in our tacit and subconscious minds (see Figure 2.4).

Knowledge is recognized as the fourth business resource alongside time, money, and people. *Knowledge management* (KM) is a natural

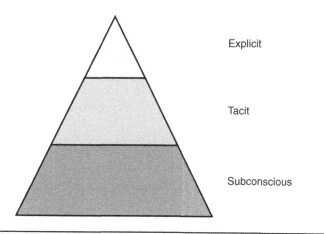

Figure 2.4   Releasing knowledge.

outcome of the work of the last 25 years that has focused on quality and process management and that developed the ideas of learning and continual improvement.

To innovate, we must grow our knowledge and develop new ideas from our knowledge. KM provides the platform for us to innovate.

HOW WE GROW KNOWLEDGE

Once you understand your information flow, you can start to grow your organizational knowledge. However, KM is more than just the flow and storage and retrieval of information. It is also about learning and growing the knowledge of the organization and using that knowledge to create new products and services. Many think this is an IT issue. To say that KM is an IT issue is to say the minds of our people contain no knowledge. Europeans see KM as very much a people issue. This is a question of balance. Balancing people and process and technology is the key to success.

Remember the types of knowledge in an organization. The knowledge inside the individual is referred to as *tacit knowledge*. The other type is the knowledge that has been documented, called *explicit knowledge*.

The *knowledge management cycle* created by Nonaka and Takeuchi (see Figure 2.5) shows the importance of the rapid transfer of tacit and explicit knowledge between people. It has four stages. It is a simple model that groups tacit and subconscious knowledge together.[5]

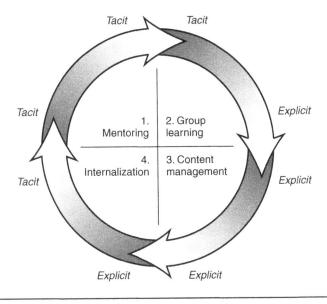

Figure 2.5 The knowledge management cycle.

Stage 1

For centuries the first method of learning has been analogous to the "sorcerer and the apprentice." The secret of success in your life was to find a good mentor. This process involved the transfer of tacit knowledge between two people. From when you were taught how to ride a bike to having lessons from a golf pro, you learned the hidden mysteries, you were "personally coached."

Stage 2

As organizations have grown, the ratio of available mentors has declined, so group learning has become important. One of the earliest recorded examples of group learning is Aristotle teaching Alexander the Great and his contemporaries. This is how Alexander was able to develop his elite officer corps. A key feature necessary for this process to succeed is *dialogue*. Knowledge is also documented during this process, and moves from tacit (stage 1) to becoming explicit. This stage of group learning is also where we introduce one of the best-known KM tools, *communities of practice*, which then morphs into *communities of innovation*. I will talk about this more in Chapter 12.

Stage 3

As the bodies of knowledge in the world have grown exponentially in the last decade, it has become necessary to store explicit knowledge in such a way that it can be easily accessed. The ancient profession of *librarian* has taken on a whole new meaning. This *content management* is a subset of knowledge management. This is where organizations have wasted mountains of money and built mountains of frustration through badly implemented IT systems. The move to overdependence on explicit (documented) knowledge has reduced the agility of organizations and the ability of organizations to innovate.

Stage 4

Explicit knowledge must be stored in a way that will enable it to create value. The core problem is not the IT system, but the method by which tacit knowledge is recorded and the method by which explicit knowledge is stored. In the past, both have been very badly executed. The objective in this last stage is for people to be able to retrieve explicit knowledge and make it become tacit knowledge. This rarely happens successfully.

COLLECTIVE KNOWLEDGE

So, to innovate successfully you need to develop linkage of processes, people, and technology, and these are primary attributes of an effective management system. The belief that technology will solve all information issues ignores the important transfer of information and decision making that occur during meetings, whether in groups or one on one.

To emphasize how the flow of knowledge and information between people takes place during meetings and in communities of innovation, let's look at a comparison of documents and meetings (see Figure 2.6). (*Bandwidth* denotes the amount of information transferred in a given time.)[6]

Taking this a step further, the book *The Wisdom of Crowds* by James Surowiecki shows how breakthroughs come not from the "genius," and how collective knowledge is so much more powerful.[7] If you go to a racecourse, the bookmakers do not set the odds based on their own knowledge and experience even though it is probably far more extensive than that of the race-goers. They rely on the "wisdom of crowds." I explain this "wisdom" more fully in Chapter 12, "Innovation Teamwork," and Chapter 14, "Networking."

	Interactivity	Bandwidth	Reusability
Document	Nil	Low	High
Meetings	Very high	Very high	Low

Figure 2.6   Transfer of tacit knowledge between people.

The challenge for an innovative organization is to release tacit and subconscious knowledge and convert it into new products and services.

The technique for doing this is *networking*, which recognizes that if you combine the tacit and subconscious knowledge of two or more people you have a powerful combination. Collective knowledge is far more powerful than the knowledge in any one individual or in any manual.

Tapping into this collective knowledge requires us to radically change many of our behaviors. We have traditionally veiled new ideas in a cloak of secrecy, we have traditionally managed our time so there is not a single wasted moment in our day, and we have increasingly focused our improvement efforts on our processes and not our products. Finally, embracing failure is something we all struggle with. If you want to innovate, all of this must change. I will talk more about this in Chapter 11, "Culture." One more thought on releasing knowledge follows.

TAKING A BATH

As with Archimedes, having a hot bath and relaxing is one of the best ways of releasing knowledge. The "Eureka!" story is of course one of the great stories in the history of physics.

Archimedes had been tasked by the King to find out whether the King's crown had been adulterated with base metals. Archimedes had struggled with the problem for a long time. Our physics teacher told us that it was through taking a bath that Archimedes discovered the concept of specific gravity. He got in the bath, the water overflowed and Archimedes captured the concept of specific gravity. He then ran naked down the street shouting "Eureka" ("I've found it"). What the teacher does not tell us is that by taking a bath and relaxing, Archimedes released his tacit and subconscious knowledge.

We know equally well that relaxing over a glass of wine, beer, or a hot drink releases many thoughts that have been locked in our subconscious.

In the modern world of business we don't take enough time to do these things. New knowledge is the fuel of innovation, and that knowledge comes from you and the people you interact with.

So the next question is "What is your own best role in this process?"

KEY POINTS

- Separating the left and right brain reduces our ability to innovate.
- The traditional hierarchical organization structure inhibits learning, while the process approach improves learning.
- A networked organization is the most effective at transmitting information and knowledge.
- Social, technical, and biological networks function in the same manner.
- For successful innovation, the internal and external networks of the organization must be developed.
- A learning organization becomes a competent organization.
- An innovative organization enables information to flow freely between its people and also enables information to flow freely to external organizations.
- An organization must have an effective quality management system to innovate successfully.
- Knowledge management (KM) is the platform from which innovation is developed.
- Documented, or explicit, knowledge accounts for only 20 percent of the knowledge in an organization.
- An innovative organization enables the rapid transfer of the knowledge in people's minds (tacit knowledge) by using its network.
- We have to allow time for the release of tacit knowledge.
- Communities of innovation are used to initiate innovation

3

The Roles and the Culture

We all have ability. The difference is how we use it.

—Stevie Wonder

When my daughter reached the age of 15, she started to talk about career options. She was very clear on her choice; she wanted to become a beautician. Well, at the age of 15 there are few girls who would not like to become a beautician. It's a bit like boys at the age of 15 wanting to become a rock star. We thought about this, and a friend of mine, Bill Rowney, told me he had taken his son to vocational guidance counseling and how happy they had been with the outcome.

So that's what we did, and the counsel we received was that because of her spatial ability and interpersonal skills, Rachel was the perfect fit for her chosen career. We sent her to a specialist college, and she qualified as a member of the Society of Aestheticians. She specialized in skin care, and after a very successful and happy career as a beautician, she went on to be a spa manager. I then think of my own career and how receiving a chemistry set as a child sent me down the road of being a chemical engineer. However, my skill sets emerged in the field of writing, speaking, and training, which is where I am now, but it took some time to get there.

So many of us struggle to find our role in life, and it is so rewarding to see someone find their perfect fit. I am reminded of the words of a friend of mine who is a counselor. Joe said to me once, "Do what you do well." Simple words, but so often we struggle in a job or career where we arrived through a set of circumstances that were not of our making. We may do well in the job, but we may not be happy in the job. Too often, we see another job and it looks attractive, so we are drawn to that type of job, but we do not have the aptitude for it.

Innovation is the same. Many people say, "I'm not creative, so I can't contribute to innovation." In truth, one of the toughest jobs in innovation is getting the new offering to market, and this requires people who are far more results oriented. Other people say, "I'm not an ideas person," but their skill may be in taking an original idea and refining it.

CULTURE

To quote the old saying from Deal and Kennedy, *culture* is "the way we do things around here." Put another way, culture is about behavior and the way we do things. As quality management developed through the '80s and '90s, we recognized that the old "macho" culture, which dealt with issues as they arose, had suffered from wide variation in the way things were done and, consequently, wide variation in what was delivered to the customer.

This led us into the process culture of the '90s and extensive documentation in an attempt to capture the knowledge in an organization and produce consistency in its deliverables. This brought a degree of order into the chaos of the macho culture but also reduced agility and reduced the ability to respond to new situations.

At the same time, we saw the emergence of the "team culture" with its focus on "synergy" and mutual support. This had a lot of appeal to business leaders and shareholders because it meant that everyone gave 110%, or to put it another way, we got "a quart out of a pint pot." This was good for the organization but not good for the individual. If people were not "team players," they were ostracized, and as we now know, the "heroism" that comes with this culture leads to inevitable burnout of the heroes.

Business now aspires to be an "agile" organization that has the flexibility of the macho culture, the reliability of the process culture, and the relationships of the team culture. There is a recognition that over 80% of an organization's knowledge is in the minds of its people as conscious (tacit) knowledge and subconscious knowledge.

A *culture of creativity and innovation* is one that releases the tacit and subconscious knowledge embedded inside its people. New behaviors are needed to release this knowledge: exploration in which we "step out of the box" and gain new learning, collaboration in which we interact with people who are "different" from ourselves, willingness to experiment and to fail in order to find new solutions. The vital spark comes from the diversity of our people.

We need to identify the right people to work on an innovation project. We need people who have the right attributes at each stage in the project, and the assessment tool in this chapter will enable both you and your colleagues to find where they make the best contribution. I will talk more about culture in Part III.

KNOWING YOUR BEST CONTRIBUTION

Figure 3.1 shows a chart that helps you assess where you will make your best contribution. A word of caution: if you score highest in, say, column four, which is the *developer* (by the way, that is my highest score), you can still contribute in other areas. In fact, you will see that it is essential to maintain a mix of all types of people as you move through the innovation process. On the other hand, it would be fatal to perform the development stage of the process without a majority of people who scored highest in that category.

One other word of caution: don't look at the column with your lowest score and rush to try and improve that score. Remember, do what you do well!

The key roles in the innovative organization and in the innovation process match the stages in the process (see Chapter 4). If you conduct this self-assessment, it will show you where you will make your best contribution to innovation, whether through:

- Generating opportunities (creators)
- Linking those ideas to solutions (connectors)
- Turning ideas into practical solutions (developers)
- Delivering solutions and getting things done (doers)

So what will be the best contribution you can make to innovation? Will it be through generating opportunities, linking those opportunities to solutions or ideas, turning ideas into practical solutions, or delivering the final solution? We all have a role to play. Score yourself on Figure 3.1.

1. Look at the first row (I like to find answers, I like to finish, I like to explore, and so on). Score four points for the phrase that *best* describes you. Then, on each row below, score four points for the phrase best describing you—creator, connector, developer, or doer.

1	I like to find answers		I like to finish		I like to explore		I like things to work	
2	I need to understand an issue		I make things work		I see both sides of an issue		There has to be a right answer	
3	Don't tell me what to do		Give me facts, not theory		I create choices		I like to analyze data	
4	I am open-minded		I convince people		I have lots of ideas		I find a weakness	
5	I "connect the dots"		I get things done		I like possibilities		I bring things "down to earth"	
6	A concept must be sound		I like "energy"		I don't fuss with details		I like precision	
7	I do not like confusion		I avoid theory		I avoid decisions		I do not like failure	
8	I think things through		I take risks		I like to hear about problems		I focus	
9	I like solutions		I like end results		I like opportunities		I like simplification	
10	I want to own the problem		I find a way that works		I like the big picture		I am thorough	
11	I like to define the problem		I want agreement		I find out the facts		I plan	
12	I want ideas		I want to try new things		I want space		I want structure	
	Total		**Total**		**Total**		**Total**	

Figure 3.1 Creator, connector, developer, doer?

2. Go back to the first row and, working down the rows again, score one point for the phrase in each row that *least* describes you.

3. Finally, score three or two on each row for each of the remaining phrases—three points for the choice that is more like you and two points for the choice that is less like you (you must have a four, a one, a three, and a two in each row)

4. Total your columns.

Row 1—you're a connector, row 2—a doer, row 3—a creator, and row 4—a developer.

Which column contains your highest score? Column 3, the *creators*, typically find the opportunity and open it up. If you do this you are likely to be an artist, marketer, or researcher. You are practical and you learn by doing.

Column 1, the *connectors*, are the "green thumbs" who nurture the seed of the idea and find the answer. They are the design, R&D, and strategic planning people. They are a rare breed and learn by thinking.

Column 4, the developers, also learn by thinking but make the idea work. They are engineers, systems developers, and accountants.

Column 2, the doers, finally take it to market and get the job finished. They are project managers, salespeople, or folks from production. Like the gold diggers, they are practical people, and like the gold diggers, they learn by doing.

WHAT DO THE SCORES MEAN?

In Chapter 12 I will explain how you form your community of innovation, and you will see that diversity is important so that ideas will be challenged. Doers and connectors are opposites, while developers think creators are not focused. On the other hand, creators think developers don't see the big picture. You can see a tension building, but this is important.

When people have completed the scoring they often ask, "I have two scores that are almost identical—what does that mean?" I can't address all scenarios—that would require one-on-one discussion—but there are a couple of common situations. First, it is clear that you can contribute in both areas of the process. However, if your scores for connector and developer were similar, it means you like to think things through, whereas if your scores are the same for developer and doer, you work best on a deadline and you thrive on time management and efficiency.

You know from your own experience that some people are good at thinking things through, while others learn from experience. In truth, our minds are more complex than this, and we are all a mix of both. For my own part, I am labeled as a *thinker*, and yet I know that I have to do something practically in order for the experience to lodge in my mind.

Let's take a closer look at each of the roles.

CREATOR

Creators learn from experience. If you are a creator, you see something happen, and the event impacts you. Your mind races on to see other opportunities related to the event you have witnessed. You see unfulfilled needs. However, your mind does not like boundaries, and you will flow from one opportunity to another. You are an observer, but you need to capture your thoughts before they move on. The skill of note taking is something you should develop, though this may be alien to you. I talk about this further in Chapter 13, "The Competent Innovator."

Because you don't like boundaries, you prefer your ideas to be "gray" at the edge. You don't like people asking you to define a problem, and you don't like people asking you to make decisions. In fact, you like to keep on generating more and more choices. You operate best in a loose mode, free of boundaries. You may have disliked those school exams where there was a "right answer" but probably enjoyed those where you could just flow with your thoughts. You love possibilities. You are an exciting person to work with, but beware of conflict with developers. They are highly focused people, and they will accuse you of "lacking focus." That's okay, that's not your job. My wife is a creator and I am a developer, and we work well together if we remember that our aptitudes lie in different areas. As a creator, you also need the opportunity to explore in order to succeed. I talk more about exploration in Chapter 13.

As a creator, you will be most comfortable in an artistic or research environment. This means graphic arts, market research, or an activity with a lot of human interaction. Ask yourself if you are doing what you do well.

CONNECTOR

The creator's best friend is probably the *connector*. If you scored equally on both, you are interchangeable, which speaks to the fact that both types work best in a loose mode. Although the name has changed, both people operate in a creative mode. However, where creators are the problem finders, connectors are the problem solvers.

The linkage between creators and connectors is important. Creators do not like to put a definition on a problem, whereas connectors do. In a large population, you will find that true connectors are in short supply. That is why people providing training in problem solving have had open season during the last 20 years. Although generally in short supply, in the world of quality management there can often be an above-average number of connectors. The quality management profession attracts problem solvers. I mean *true* problem solvers.

Most people jump to a solution and implement. The danger in the world of the innovator is that the creator sees an opportunity, and we implement the first solution that springs to mind. Connectors don't do that. Importantly, they define the problem the creators have uncovered and, even more importantly, "connect" to many solutions. Connectors will take solutions from a different context and put them into the context of the current problem, but keep looking for better solutions. A good example of this was when Henry Ford was seeking the solution to his problem of mass production and saw a meat processing plant where the carcasses were being moved by hanging them on hooks; for years after that you saw car parts moving around a factory on hooks.

This "connecting" happens to all of us at some time, and it often happens so quickly we think it is "magic." In truth, connectors may work on a problem for a long time before they find a solution. The danger, therefore, is in stopping at this one solution; to get the best solution, we need choices. Connectors love problem solving, and so they need a diet of problems. If you are a connector, compared with creators you are far more of a thinker. Because you are a thinker, you will really want to understand the problem and you will address solutions from a conceptual rather than a practical level. However, you will want your concept to be sound.

Because you are a thinker, you may not be involved in implementing the solutions. Don't be insulted, that is not your job. Linus Pauling, the Nobel Prize winner, said the best way to have a good idea is to have a lot of ideas—and throw away the bad ones.

This is the job for connectors. One word of warning: Traditional problem solving has been very left-brain and analytical and tends to find solutions within the existing context. The innovator looks for radical solutions. This means using the right brain. I will talk more about this in Chapter 7.

SELECTING THE SOLUTION

Selecting the preferred solution is something I will also discuss later. This is the job of the strategic planners. The connectors need to provide them with choices, and those choices need supporting data on risk, ROI, and practicality.

Many people think that once you reach this point, innovation is finished. What I have described so far is not *innovation*, it is *creativity*. Creativity is a subset of innovation.

Innovation is about converting new knowledge into new products and new services. So far, we have only created new knowledge. We do not yet have the new product or service. Now the game changes, and we go from "loose" to "tight." We have to make the solution practical, and we have to get it into the hands of the user.

DEVELOPERS

The *developer* is far more focused than the connector and the creator. If you are this person, you will work best in a defined or project-driven environment. You want a specific problem to work on, and that problem must not be ambiguous. The many ideas from the connector must have been distilled into two or at the most three choices to enable you to focus. You are similar to the connectors in that you are both thinkers, and so you need to take time to understand the concept they have developed. Your strength is that you turn abstract ideas into solutions that work. You are good at data analysis and so can pinpoint weaknesses in a product or a process and move on to the best solution.

Time is of the essence at this stage in the innovation process, and you might find this problematic because you want your solution to be precise and unambiguous. You hate being taken off the job you are working on before it is complete and being asked to work on something else. You will miss lunch or dinner if you have in your mind a completion point for something you are working on.

This is the attribute where I have the highest score, and I envy those people who can leave off on a task that is incomplete, relax and enjoy lunch, and then come back to the job later. Taking time to relax is a skill that developers have to work on. This is when developers can have some of their best insights. Remember that Archimedes found his answer to specific gravity when he took a bath. A lunch "out of the box"—and I don't mean a "boxed lunch," I mean out of the building—can often be your best friend when you are hitting the wall and not finding answers.

Because you are focused, you can find yourself in conflict with the creators, who like to see the big picture, and you can also forget to include the potential users of your solution. Don't forget the "alpha testing" that is fundamental for the developer. You are probably a good planner, so use that skill to create your project timeline and also to involve the implementers, or doers, who come next.

DOERS

The *doers*, like the developers, work best in a project-driven environment and need the tight mode of operation. You may think that if you scored highest here, you have no job in the world of innovation. In truth, your job is one of the toughest of all. You need to have been involved with the developers so you know what is coming down the road. You are similar in that you work best in a structured environment, but you are different in that developers are thinkers, whereas you like to "do," as your title suggests. Be aware of this and respect your differences. You may be surprised to find that you have strong similarities to the creators in that you are both practical people. You both enjoy new situations and find them stimulating. You can be impatient, and, unlike the developers, you will keep "breaking things," trying different solutions until you find something that works. Use the thinking strengths of the developer to help you here. When you are getting your feet wet and your hands dirty, you feel you are getting something done.

You are likely to be in operations or sales, if you are in your natural habitat, because you like to deliver. Because of your practical approach to life, if you are in operations, you find to your surprise that you mix easily with the sales and marketing people, but you are less at ease with those R&D folks. If this is the case, be conscious of it and learn to work with those "thinkers."

Your challenge is to get the solution in the hands of the customer. If you're an operations person, this means working with the developers to eliminate production or service problems. If you are a sales person, you need to understand the opportunities found by the creators in marketing who originally saw the opportunity.

LEARN TO WORK WITH OTHERS

You can see that whether you are a creator, connector, developer, or doer, this does not put you in a box. You need to work with people that have other attributes in order to do your own job. Understanding how those other people think and act will help you to work better with them. Use the assessment on your team, and understand your colleagues.

These revelations may not be new to you, but they are indicators of where and how you will make your best contribution to the multifaceted world of innovation.

For success, you need a mix of people at each stage in the process. The tension is healthy!

One other thought: Your work environment will, over time, change your personal attributes. As I have chaired the ASQ Innovation Division, the task of "getting results" has increased my score as a "doer."

So, having identified our role in the organization, what is the process we must follow? How do we bring this mix of people together to innovate?

KEY POINTS

- We all have a role in the innovation process.
- Creators generate opportunities.
- Connectors link opportunities to solutions.
- Developers make solutions practical.
- Doers implement solutions.
- A self-assessment can show you and your colleagues where you will make your best contributions to innovation.
- Creators and doers are practical.
- Connectors and developers are thinkers.
- It is essential to have a mix of different attributes at each stage in the innovation process.
- The mix can create tension, but this is healthy.

4

The Process and the System

The best way to have a good idea is to have a lot of ideas—and throw away the bad ones.

—Linus Pauling

Some friends from England, Phil and Liz, wrote to us one year and suggested that for the summer vacation we do a "house swap." It reminded my wife and myself of the movie *The Holiday* in which Cameron Diaz house-swapped with a person from England, and we smiled. Then we thought, what a great idea! They would live in our house in the country in Canada for two weeks, and we would live in their apartment by the sea in England.

We agreed on the dates with our friends, made the travel plans, and then about two months before the swap it was a case of "reality bites." Our two friends were very tidy in their habits, but we were a couple who lived life "on the run" with the attendant trail of debris.

About two years previous to this, we had borrowed a book on getting rid of clutter, and that suddenly became our bible during the weeks before our friends arrived. We discovered things we didn't know we had. We dumped things we hadn't used for years, and we transformed the house into a place where there was not only room to store things, but we could also find them! It was a great feeling.

My garage had been like the world of the mad inventor: old tools, bits and pieces of metal and wood, old garden furniture, and a host of items that I was going to repair some day when I could locate the tools I needed to do the repair. I emptied everything onto the driveway, hosed out the garage, and returned to the garage only the things I really needed. The rest I dumped.

My point? Your innovation process, even in its early stages, needs to be free of junk. Your ideas need to be carefully stored and well organized like any good garage. The world of the engineer and the inventor is often like my garage before it was de-junked. We keep things because "they might come in useful." This applies to both documents and materials and components. After a while, we don't know what we have and we can't find what we want.

This happens due to the false premise that technology drives innovation. It happens through failing to understand the innovation process. Too often, the inventor creates something that "might be useful" and then spends forever looking for a problem that their solution might solve. They think innovation is a solution looking for a problem.

Occasionally, this works, and of course these are the stories that make good newspaper copy. The classic story here is of course the Post-It note.

In the early 1970s, Art Fry was in search of a bookmark for his church hymnal. Fry's colleague at 3M, Spencer Silver, had developed a failed adhesive that peeled off. Fry used some of Silver's adhesive on the edge of a piece of paper. His hymnal problem was solved! Fry saw that his bookmark could have other uses, and the Post-It note took off.

The other story, in some ways more impressive, is IBM's alphaWorks. IBM is a great innovator. Their "garage" was so full of great inventions that were not being used that the development staff was angry and frustrated. So they had a "garage sale." They put all their unused ideas on a website for free download for 90 days. Forty percent of their ideas were taken up, and they now have a multi billion dollar a year business selling previously unused products.

These stories make good reading, but they are not good innovation stories. The truth is that the great inventions are driven by need and not technology, and that the need is not always recognized. Yes, the second stage of the innovation process may generate solutions that could work in a different context. Yes, there will be many failures before the final success. However, if you clutter your life with your failures, your innovation will grind to a halt. Learn from the failure and move on. A good maxim from the world of knowledge management is that you are unlikely to use explicit (documented) knowledge that is more than two years old. It will either have become tacit or have become out of date.

Innovation does not start with finding a home for a cool idea, it starts with a market need and seeing a market opportunity.

To innovate successfully, you must always be thinking about tomorrow's customer and what their needs will be. Much of quality management has tended to focus on today's customer and address their immediate concerns. When quality management has been inwardly focused it has often caused us to neglect the future. The driver for the innovation process is not always evident, and that is why people sometimes believe there is no process. The driver is tomorrow's customer, and, of course, that person does not yet exist. Phil Crosby said, "All work is a process," and yes, innovation is a process. However, it is not linear; it is a looping process. It also changes mode from loose to tight and back to loose.

The reason most people struggle with the innovation process is that it is unusual. To innovate successfully, an organization needs to manage a paradox: The innovation process operates in two distinct modes. The creative phase where we find the opportunity and the conceptual solution requires a "loose" process, and behaviors are loose and free to allow free thought and creativity. The execution phase, on the other hand, requires a tightly managed process where focus is essential and time is of the essence. Clearly a different set of behaviors. The cultural change will be in changing behavior from loose to tight. This is not easy, and I will explain this in Chapter 8.

These stages in the innovation process also point to the key roles in the process.

The idea of the lone genius is a common myth about innovation. Most innovations are the result of collective knowledge. You can see by comparing Figures 4.1 and 4.2 that the stages in the process match the roles in the process. People with different attributes are needed at each stage:

- Creators are needed to generate opportunities

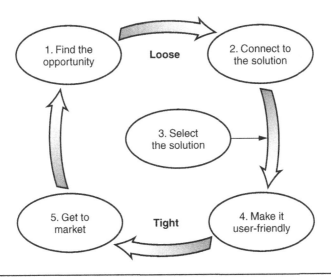

Figure 4.1 The stages of the innovation process.

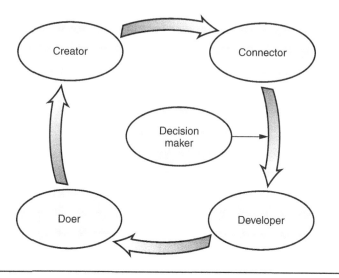

Figure 4.2 The innovation cycle—the roles.

- Connectors are needed to link those opportunities to solutions
- Developers are required to turn ideas into practical solutions
- Doers have the job of delivering solutions and getting things done

If you conduct the self-assessment in the previous chapter, it will show you which of these categories you fit into and the stage where you will make your best contribution to the innovation process.

THE OPPORTUNITY

The first step in the innovation process is to identify the opportunity. Importantly, the innovator must understand that neither the customer nor the market may recognize that need or opportunity. Henry Ford said, "If I had asked people what they wanted, they would have said 'faster horses.'" Creative thinking is usually required at this first stage, and the creators are the primary influence at this stage. They are the "gold diggers" who typically dig out the opportunity and open it up. These people operate best in a loose environment.

Typical examples of such opportunities include: IKEA found furniture stores overpopulated with aggressive sales staff and providing a confusion of choices; Southwest Airlines found that travelers were annoyed with bad meals and the large amount of leg room given to first class passengers; Yellow Tail Shiraz found that casual wine drinkers were intimidated by "antique" labels; and Cirque du Soleil found that there was increasing concern about the treatment of animals in circuses.

Ironically, in today's technology-driven world every technological advance creates new opportunity. The cell phone was created, and people needed to use it in their cars, so "hands free" devices were needed. Computer screens took up too much space, so flat screens were created. Laptops were heavy, so tablets were created. The list goes on. There is always a new need.

THE SOLUTION

Next, you find the idea that solves the problem. Once the customer opportunity has been found, then the problem solving begins. Beware, however: people often rush from "concept" to "deliver." People often rush to market with a half-baked idea. There are vital interim stages in the innovation process. A good problem definition is essential, and connectors do this far better than creators.

Radical new solutions are where most people recognize innovation as taking place. Breakthrough innovation comes from finding radical solutions. This stage is where connecting a product or process from a totally different environment often leads to that "Eureka!" moment. This is exciting! Innovation is the use of new knowledge to create new products and new services. This second stage is the application of that new knowledge. Again referring to Henry Ford, his ideas for mass production came from seeing a meat processing factory and applying its methods to the production of motor vehicles. Connectors, like creators, need a loose mode of operation.

This will often mean removing what have been regarded in the past as sacred attributes of your offering. The companies I mentioned all had the courage to do this. They all took out what were previously considered sacred attributes in their respective industries.

THE TIPPING POINT

After the second stage, we only have a concept. We then reach a tipping point where decisions have to be made. We make decisions on which is the best concept to pursue based on factors such as risk and behavior change. We then narrow our focus. This is also where our activity changes from the loose mode of the first two stages to a tight mode where the developers take over. Statistics show that out of 3000 ideas, only one will come to fruition. Unfortunately, we often kill the best ideas with overaggressive requirements for ROI. Or we let a high-risk product through because we failed to assess the risk attached to a potential new product.

Good decisions require good data, so it is essential that the connectors provide good data to the strategic planners, who make the decision on which options to pursue. We now have to develop a working solution.

DEVELOPMENT

The development stage should be fast and not secretive, and should involve the ultimate user. This is where you make the product functional and user-friendly, and eliminate the glitches. Good project monitoring and control are vital. This is where behavior changes from loose to tight. This stage in the process is where so many organizations lose momentum and lose the advantage they gained in stages one and two. Speed to market is essential. Discipline becomes vital. This is where Edison's "Genius is one percent inspiration and ninety-nine percent perspiration" comes in to play. The developers make the idea better and make it work. They are engineers, systems developers, and accountants. Many companies go wrong here by staying in a loose mode of operation. It is vital to go tight.

DELIVERY

Finally, delivery is where the operations and sales people take ownership; but remember, they must be involved in the previous stage so that there are no surprises here. Production problems must be eliminated during development. The value proposition that the sales people will use is also created during the last steps of development! Getting to market is where far too many stumble. The production, service delivery, and sales people need to have been involved in the earlier stages if you want a long-term and continuous innovation process. They now have to "run for the line," so a tight mode continues to be vital. Every advantage we can give them is essential. Time is their biggest advantage of all. The doers deliver and get the job finished.

We will look at these stages in more detail in Chapters 6 through 10. You need to know how good your organization is at each stage in this process. Assess your process at the end of each of the chapters!

I have given you a core process for innovation, but it must sit within a management system if it is going to succeed.

THE QUALITY MANAGEMENT SYSTEM AND INNOVATION

Peter Senge's book *The Fifth Discipline* quotes the famous statement "The only competitive edge an organization has is the ability to learn faster than the competition."[1] I will extend this to say, "The only competitive edge an organization has is the ability to create and apply new knowledge faster than the competition." To achieve this flow of information and knowledge and become a learning organization, the organization's processes need to be networked to come together as a system.

In a successful organization, the business processes link to form a management system, and this enables the free flow of information, which in turn leads to the creation of new knowledge. Knowledge feeds innovation, and innovation feeds success.

An effective *quality management system* (QMS) is an essential delivery system for the innovator. You can come up with the best new idea in the world, but if you can't deliver, the competition will jump in and fill the vacuum of demand you have created. The most widely used QMS is ISO 9001. However, remember that the QMS focus is on delivery, and also that a poorly implemented QMS will kill individual attempts at creativity. The most symptomatic illustration of this QMS problem is the mindless "say what you do—do what you say" axiom that persists, and yet ISO 9001 is about improvement driven by measurable objectives, not blindly following procedures.

The primary system-level obstacles to creativity in ISO 9001 lie in these areas:

1. Overdocumentation
2. Management review without influence
3. Internal auditing simply for compliance
4. Corrective and preventive action at a low level
5. Failure to engage people in the QMS

The first step in addressing system obstacles is to clean house. You have almost certainly invested in technology, and you have probably invested in developing the competence of your people. There is a fair chance that a proportion of the documentation you created originally has become redundant. In addition, question the value of each remaining documented procedure. An effective quality management system is lean and flexible.

For innovation to succeed, it has to occur within a management framework, but one that is agile. If you are from a software background, you may have encountered the "Agile Manifesto." It describes how a successful organization has a management framework that values improvement over compliance, lean over artifacts, learning over habit, and autonomy over authority. In each case, one does not exclude the other. It is a question of balance.

A good QMS engages the people in your organization, and it produces nonfinancial data, which in turn produces knowledge you can use for innovation of both your processes and your products. Taking this a step further, innovation management must fit with quality management. As an indication of the acceptance and importance of innovation, we are seeing the development of national and international management system standards. For example, Ireland developed one of the first national standards on innovation back in 2009 (see sidebar). They wrote their standard using the model described in the first edition and current edition of this book. Naturally, they adopted some new terms.

IRELAND NWAI 1

1. Investigate
2. Choose opportunities
3. Generate solutions
4. Choose solutions
   - Prototype opportunities
   - Build/test
5. Develop solutions
   - Design spec and scoping
   - Build business case
   - Design and develop
   - Test and validate
   - Stop and store what you have
6. Produce
7. Launch and review

THE INNOVATION MANAGEMENT SYSTEM

There are three levels to innovation, just as there are with quality. They are *system*, *process*, and *product*. For *system* you can read *organization*; for *product* you can read *tangible* or *intangible*. Delivery of intangible product means the delivery of a service or of knowledge.

The IBM Global CEO Study,[2] in their report *Expanding the Innovation Horizon*, showed convincingly that organizations who innovated at a system or organizational level grew operating margin more rapidly than the competition. This is not surprising to people experienced in quality management and who know, to quote Deming, that 90 percent of the problems in a process are the result of the system in which the process operates. In turn, the way you get your product right is by focusing on the process that produces it.

ISO/TC 279 on innovation management was tasked in 2014 with developing an innovation management system (IMS)—ISO 50500. The structure is the same as ISO 9001 and follows the same high-level structure adopted by ISO for all management system standards. As we move to integrated management systems, the IMS and quality management system must be clearly linked.

An *innovation management system* is fundamental to successful innovation. An innovation process is fundamental to the creation of new products and services.

This doesn't mean you only innovate at one level; you must innovate at all three levels: system, process, and product. If you don't get your products, services, and markets right, the other stuff doesn't matter. Product innovators also outperform their competition on operating margin.

I will discuss management systems more fully in Chapter 16.

This brings us to the strategic decisions we must make as innovators.

KEY POINTS

- The innovation process must be kept tidy and organized.
- Innovation is driven by need, not technology.
- The early stages of the innovation process operate in a "loose" mode, the later stages in a "tight" mode.

Continued

Continued

- The opportunity stage is where we find a market need that is not satisfied.
- The market may not realize the need exists.
- The solution stage is where "connecting" to a different environment often reveals a radical solution.
- The best way to have a good idea is to have a lot of ideas (solutions).
- A preferred solution is selected based on time, cost, and risk.
- The solution must then be made user-friendly to enable adoption.
- Delivering the working solution is one of the toughest challenges.
- Only one out of 3000 ideas "makes it."
- An effective quality management system is essential to enable delivery at the final stage of innovation.
- An innovation management system must integrate with the quality management system of the organization.

5
Innovation Strategy

If you see a bandwagon . . . it's too late.

—James Goldsmith

Business is about competing. Or is it? We are so focused on the competition that we miss new opportunities that are staring us right in the face.

My friends tell me I am too competitive, and in many ways this is because I was raised in an education system where unless you were first, either academically or in sport, you had failed. As a result, competition was in my blood, but it seemed I was always trying to be first and not quite making it.

Twenty years after leaving my school, King Edward VI School in Aston, Birmingham, the school had its 100th anniversary reunion at Penn's Hall in Sutton, Coldfield. It was early evening, and I was at the bar with my friend Bob, ordering drinks for us and our wives. Nearby were a couple of guys a little younger than us, and one looked at me and said, "Aren't you Peter Merrill?" I looked at him a little surprised, as I didn't recognize him, and simply said yes. "Wow," he said, "you're Merrill, the runner! We used to think you were a god!" I was totally stunned. This was a new reality. After so many years as seeing myself as an average sportsman, here was someone who saw me as a great competitor.

It made me realize that in our striving to compete we miss so much that is right there in front of us. Competition is not always good. It can blind you to reality and to opportunities on your own doorstep. Most businesses focus on beating their competition, and they compete on small cost and quality improvements to existing products. They overlook the fact that customers are making major substitutions for their product by purchasing radical alternatives from other industries. Partygoers stopped buying wine and instead bought vodka and cranberry juice. People stopped going to the circus and started going to Disney. These are simple substitutions. Many substitutions are more extreme.

In business we are obsessed with competing, and yet the perfect strategy is to find a market where there is no competition by creating an offering that is unique and that no one can copy.

MEGA TRENDS

In pursuing this strategy you also need to be aware of mega trends that are creating new demand. Technology, the environment, and health are examples of these influences. At any industry or business level there will be specific influences.

To understand mega trends, look back in history and ask yourself what the eras of the last two thousand years were remembered for. "The glory that was Rome" was clearly about architecture and literature. When civilization emerged from the Dark Ages in the 1500s, we think again of architecture and literature, but fashion and clothing also become significant. In the twentieth century we see the effects of the Industrial Revolution and mechanization: planes, trains, and automobiles, which were really about communication. Today, we are seeing the information age and technology being used to store and transfer information, or what we call IT (information technology).

However, other things are happening as a consequence of these advances in civilization, and a lot of them are negative. The chemical industry has a lot to answer for as a result of its behavior in the twentieth century. I look back with disgust at the total lack of values, in fact destruction of values, in my education at the Chemical Engineering School in the University of Birmingham. Today's environmental movement to "save the planet" is addressing one hundred years of major abuse by the chemical industry.

A second major trend is related to this: the "social responsibility" movement. It's a consequence of improved communication. Globalization has led to people in the world's poorest countries being well aware of the lives of the "rich and famous." Equally, people in the developed world are becoming more conscious of exploitation in the third world. This fuels the equalization of wealth and affects issues like global terrorism.

Futurism has been around for a while, and I do not intend to address it in this book. But one should be aware of it, and also aware of the impact of future trends on your business. Books like *Megatrends* by Nesbitt[1] and work such as that conducted by the International Organization for Standardization

(ISO) will point you toward high-level trends. Read *The Limits to Growth* by the Club of Rome,[2] another book that has shown an eerie accuracy in its predictions of the future.

YOUR OWN MARKET OPPORTUNITY

The question you have to ask is what is the next level of "trending" in your own market, and more importantly for yourself, where are people in your market having difficulty in "getting something done"? It is easier to create a new product or service for the market you already serve than to create a new customer base. Look at your own market space first. You need to learn what changes are affecting your customer, both obviously and subtly, and the emotional effect those changes are having.

Companies conduct what is called an *environmental scan* to identify these trends and their effects on their customers. Change at a strategic level is affecting your customers, and it is essential to identify those changes before you talk to your customer. You need to know the legislation, the technology, the foreign competition that is affecting your customer—the environmental scan, which should be a standard practice in marketing. The scan is a technique for identifying where changes in the market environment have created "pain" for the customer. The forces causing this change will be economic, social, political, low-cost economies, and the environment. The output of this activity is the *customer pain statement*. Techniques for finding this pain are activities such as focus groups and "pain storming."

The accomplished innovator then looks for what is sometimes called *white space* or *blue ocean*.[3] I prefer *green field*, which is an established business term that means "go to an uncontested market where there is no competition." You create a new demand and you provide a better product or service, also at lower cost.

In the 1980s Ralph Lauren recognized that the fashion market had split into the couture names, such as Armani, Fendi, and Versace, and the commodity brands such as Gap and Hilfiger. The top end had gone glitzy, and the bottom end had gone drab but pricey.

Lauren built its Polo brand in the green field between the two market groups. It lacks creativity in design and charges high prices, but it has a designer name, elegant stores, and most importantly, exquisite fabrics. Garments lacking fashion but made with quality fabric became its strength, and at a lower price than, say, Burberry. Polo's timing was perfect to meet the business casual market of the '90s.

If you have a polo player embroidered on your golf shirt, you can wear it to the office and everyone knows you must be a smart person. People have tried to copy the success of the brand, but the cachet of Polo has been maintained through excellent marketing.

More recently, the financial crash of 2008 saw people tight for cash. This "social change" made people interested in saving money. Groupon launched in November 2008, and the first market for Groupon was Chicago, followed soon after by Boston, New York City, and Toronto. By October 2010 Groupon served more than 150 markets. However, the concept of the discount coupon was easy to copy, and in the UK the Brand Leader is Vouchercloud, a money-saving mobile app providing consumers with vouchers from a range of categories, including restaurants, travel, leisure, and hotels. Vouchers can be redeemed through online voucher codes, printable vouchers, or through the mobile app. Parent company Invitation Digital Ltd was also set up in 2008 to deliver digital incentive promotions and loyalty schemes across Europe.

THE FIVE FORCES

In defining where we will direct our strategy, Michael Porter's *five forces* model is the original model for business strategy.[4] He sees five competitive forces that shape every industry (see Figure 5.1):

1. Bargaining power of suppliers
2. Bargaining power of customers
3. Threat of new entrants
4. Threat of substitutes
5. Competitive rivalry between existing players

Porter argues that the key to sustainable competitive advantage is either *cost advantage* or *product differentiation*, both of which relate back to the five forces model. He included a third factor, which is to be a *niche operator*, but even this will be either on a product cost or differentiation basis. His conclusion is that strategy is about choosing one route over another.

Too often in their obsession with being competitive, businesses slug it out on cost or on quality.

With globalization, competing on a cost basis is becoming less and less attractive. Porter, in his book *Creating and Sustaining Superior Performance by Competitive Advantage*, lists the following factors as affecting competitiveness.[5] I give you a typical example in each case:

1. *Entry barriers*. High capital investment and time to build, such as entry into heavy chemicals.

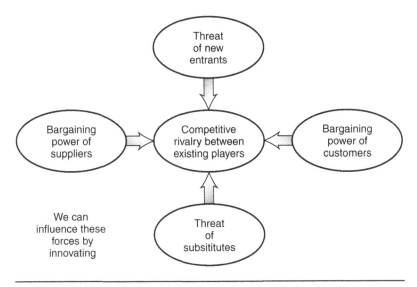

Figure 5.1 Competitive forces affecting survival of a business.

2. *Supplier power.* The oil industry suffers from raw material being located only in certain areas of the planet.

3. *Rivalry factors.* The retail industry is easy to enter and so is incredibly competitive.

4. *Substitution threat.* This is often subtle, and the telecom industry, which for decades had a monopoly, now has to deal with this in a major way.

5. *Customer power.* People can walk into and out of a restaurant at will.

At the same time, you need to remember to "do what you do well." In his book *Good to Great*, Jim Collins talks about the *hedgehog principle* and how successful companies pursue what they are passionate about and also do what they can be "best in the world" at doing (see Figure 5.2).[6] He also stressed that, at some point, you have to face the brutal truth that today's product is fast becoming obsolete and that you need to "innovate or die!" This means use your best abilities but point your abilities elsewhere.

In many ways, Porter's work has been superseded by Rita McGrath, who shows us that competitive advantage is only transient given the speed of change.[7]

Transient advantage means companies in high-speed industries must learn to move fast through the stages of competitive advantage. They also

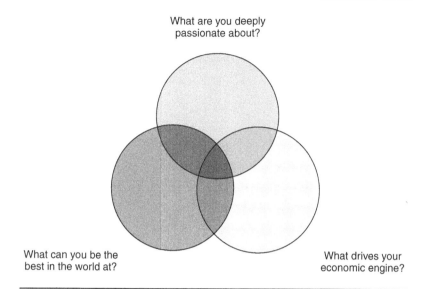

Figure 5.2 The hedgehog principle.

need the capacity to develop and manage a pipeline of initiatives, since many will be short-lived. Her model is thus as shown in Figure 5.3.

McGrath then stresses the need to think about market areas, not industries. Nontraditional competitors always take companies by surprise. Google's moves into phone operating systems upset the traditional phone companies; retailers like Walmart moved into healthcare. Outline the market space you would like to claim, and then let people experiment in that space. Be prepared to make a shift as new discoveries happen. Google did this well. Be systematic about the opportunity stage of innovation: advantages disappear, so have a process for filling your pipeline with new opportunities.

These points were all stressed in the first edition of this book, and are just as true today.

START WITH YOUR OWN CUSTOMERS

Too often, people will work on something they find exciting but that has no market relevance. More commonly they follow convention. Convention says compete in an existing market, beat the competition, serve existing demand, and provide either a better product or lower cost.

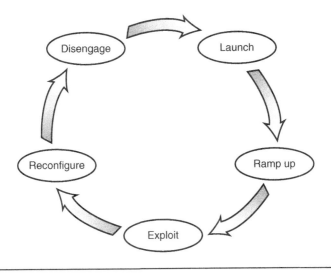

Figure 5.3   Launch, ramp up, exploit, reconfigure, disengage.

To initiate innovation strategy, you begin with your customer but talk about the customer's needs . . . not yours. What are they having trouble getting done? Work with your most demanding customers and include sales as well as R&D in the discussion. You need to know what would make their lives easier.

GE had a major market in the airline industry. Soon after 9/11 they moved into security equipment that outdated the metal scanner. Soon after the financial crash of 2008 Groupon saw that people were tight for cash and developed their coupon discount business. Cirque du Soleil moved from the dying circus market to theatre, but still did the amazing acts. Milliken, who had been the benchmark in chemicals and textiles, is now a world leader in advanced materials and left textiles in the '90s. They never stopped innovating. Do what you do well, but pick your market carefully.

You need to know the issues your customers dislike or that upset them. Those issues are always changing. To get a good sense of what this means, make your own "hassle" list. In fact, make two lists, personal and business. Here are my two lists of pet hassles, in no particular order, to get you started (Figure 5.4).

Now use Figure 5.5 to make your own lists to describe your own reality. Off the top of your head, list the things that have bugged you in recent weeks or months. One list is for your personal life and the other for your business life.

Business "hassles"	Personal "hassles"
1. E-mail crashes	1. Waiting in line at a store
2. Late starts to meetings	2. Slow restaurant service
3. Airport security	3. Waiting at the doctor's
4. Cancelled flights	4. Being stuck in traffic
5. Aggressive people	5. House cleaning
6. Shock hotel bills	6. A bad night on television
7. Declined Visa card	7. Noisy people in a book store
8. Boring conferences	8. Clearing snow
9. Technical support by phone	9. Power outages
10. Filing papers	10. Doing taxes

Figure 5.4 Author's business and personal "hassle" lists.

Business "hassle"	Personal "hassle"
1.	1.
2.	2.
3.	3.
4.	4.
5.	5.
6.	6.
7.	7.
8.	8.
9.	9.
10.	10.

Figure 5.5 Reader's business and personal "hassle" lists.

As you look at your lists you can probably see that someone provides some kind of solution to a number of your problems. Some of the solutions are expensive, so you haven't adopted them; others are difficult to use, so you can't be bothered. All of these problems are innovator's opportunities, and all of the solutions you have chosen not to adopt are innovator's opportunities.

You need to be asking the selfsame questions of your customers. What upsets them?

If you have problems with power outages, you can buy a home generator for $5000, but you may prefer to buy candles and a deck of cards. That might be a good enough substitute. But if someone offered you a generator for $500, you might jump at it. You probably carry a backup credit card to avoid those "bad moments," and you probably ask for your hotel bill the night before you check out; in fact, most hotels have "innovated" their service to do this anyway, and also send the bill to you electronically. We see increasing use of kiosks or ATM-type equipment to avoid queuing, but many of us don't like these because they are often not user-friendly.

It's the organizational problems that are the most frustrating: airport security, the appalling service from phone companies, or being stuck in traffic. These are instances where you feel you "can't fight the system."

You need to know where your customers feel they can't fight the system. It may point to a need to change your existing process. Unfortunately, well over 80 percent of so-called innovations fall into this category of minor process improvements. On the other hand, if you diligently use your innovation process, it may lead to a radical new offering.

Barely 15 percent, or one in six, of so-called innovations can be described as "breakthrough" or "radical."

DECIDING WHERE NOT TO PLAY

Remember, in setting strategy you need also to decide "where not to play." This comes from staying with what you do well and using your core competencies. There may also be ethical issues such as gambling, environment, or health that influence where you "do not play." You can choose to play in whichever market works for you, but again remember, it is much easier to develop a new product than develop a new market. Also remember that too much choice can be overwhelming, so determine which markets and products the company will not address. This gives a great sense of clarity as you clear away the "clutter."

In these early stages, as you develop your innovation strategy, you need to move away, but not too far, from the core business. You may want to look at the six to 10 most successful innovations in your market's history, and also the six to 10 biggest failures. This can give you a good sense of direction. You then need to step out of your comfort zone.

This is OK for getting started as an innovator, but there are limits to today's customer's ideas and opportunities. There are more ideas outside the box. Open market innovation is your ultimate destination, which I will explain in Chapter 18.

Setting direction is done as an oversight to the first step of the innovation process, which is about creating the opportunity.

KEY POINTS

- The perfect strategy is to find a market where there is no competition.
- In setting strategy, we should note the high-level or mega trends that affect our existing and potential market.
- Understanding your market means understanding the mind of the customer.
- You find out where your customer is having difficulty; you do not just offer them your latest "cool idea."
- The customer is less interested in your actual product or service and more interested in what it will do to improve life for them.
- The "five forces" model of Michael Porter enables an organization to assess the impact of suppliers, customers, competitors, new entrants, and substitutes.
- Rita McGrath stresses that competitive advantage is only transient.
- Successful companies pursue what they are passionate about and what they can be "best in the world" at doing.
- Innovation strategy starts with finding opportunities with your existing customers.
- It is just as important to decide "where not to play" when selecting your target market.

Part II
The Process

6
Seeing the Opportunity

Necessity is the mother of invention.

—Plato

I was waiting at the departure gate at Tampa airport and started browsing through the magazine stand. A copy of *Scientific American* caught my attention. In my engineering days it had been a regular read, and I recalled the long lapse of time since I last read the journal.

I bought a copy and during my fight I skimmed through the articles until I came to an article written by Michael Webber, "How to Make Our Food System More Energy Efficient." He had my attention. Webber's article was packed with data presented in an easy-to-read and logical form, but the bottom line of his argument was that we use 10 units of fossil energy to produce one unit of fuel energy.[1] His argument in BTUs had captured my chemical engineering mind. He had also captured my innovation mind. What an innovative opportunity!

Opportunity is the start of the innovation process, and from Weber's 10:1 ratio, the food industry has an enormous opportunity for innovation. The inefficiency of converting sunlight to plant growth to animal feed to human consumption presents the innovator with an opportunity that, to use food language, should make the innovator salivate. To quote the fortune cookie saying, "The secret of a good opportunity is recognizing it!"

You may well ask, "How can I possibly take something at such a high level and make it practical?" Well, this is an example of a mega trend, and it translates in that industry into taking a more ethical and less short-term view on innovation. Michael Webber's article in *Scientific American* describes some of the really interesting innovations the industry is now starting to employ: drip irrigation, which uses 40% less water and 15% less energy; no-till agriculture, which places seeds directly into soil, reducing energy usage, erosion, and carbon emission; and laser-leveled fields, which again reduce erosion but also reduce fertilizer runoff.[2] However, that is jumping into solutions. Let me stay with opportunity.

STAYING LOOSE

The first stage in the innovation process is seeing the opportunity. People think innovation opportunities arise randomly. In truth, when you hear the word "random" it usually means, "I don't understand." This is a proactive activity, and good opportunities are not hard to find. Take a simple need: a person would like a picture of themselves. Two hundred years ago they would have employed an artist. The process was lengthy and also expensive. However, the person was involved in the process, and the picture could be adjusted during the process.

One hundred years ago photographic film provided a cheaper and sometimes quicker solution. However, film could take up to two weeks to develop, and the result could not be adjusted. The subject was not involved in the process. Attempts to speed up the process included Polaroid and 24-hour film services.

Today, digital photography enables the subject to be involved and for a picture to be retaken at zero cost if it does not meet a person's immediate need. We have fulfilled their needs. Kodak had that opportunity and declined, as did Sony, who failed to take up the opportunity of music in digital format.

I was previously a member of the Innovation Council of the Conference Board, and we met Cirque du Soleil at their Montreal base. Theirs is a remarkable story of innovation. Guy Laliberté and his friend were street performers in Montreal witnessing the demise of the circus as we knew it. However, they saw a great new opportunity for more-sophisticated entertainment for family and business. Cirque du Soleil was conceived. It's only when you see behind the scenes that you recognize how they have created a market position that is now very difficult to attack without a huge investment of not just money but time. It has taken Cirque 20 years to reach this position. They have also made an enormous investment in their artists, their training techniques, their design methods, and their relentless innovation.

It is interesting to compare Cirque with Yellow Tail. Yellow Tail was the brand leader in its party niche, but their niche was entered by a veritable menagerie of competition, including Long Neck from South Africa and Little Penguin from Australia, to name just two. It has been relatively easy for competition to enter this new market. All you need is the product analysis and then you can copy it, especially in a cheap wine.

There is a key lesson here for the innovator. People can copy your offering, even steal your technology, but it is much harder for them to steal your unique competencies. Always try to innovate in your area of greatest competency.

I mentioned earlier how the "social change" of the financial crash of 2008 saw people tight for cash, and Groupon launched in November 2008. However, the concept of the discount coupon was easy to copy, and in the UK the brand leader is Vouchercloud. It is unclear who copied who since they both launched in 2008.

The innovator looks for what is often called *green field*. You create a new demand, and you provide a better product or service at lower cost. You need to be looking for that perfect market where no competition exists and others will find it hard to enter. You never stop innovating!

You must understand the market you are in. What does your customer really need? The customers of pharmaceutical companies do not need drugs. They want to avoid being ill. In the Third World the competitor of the pharmaceutical company is an organization like WaterCan, now called WaterAid, which is developing clean water supplies for the world's poor.

You must be watching trends and seeing the big picture, finding untouched markets that are creating new demand. But at the same time there will be specific influences at any industry or business level.

This "opportunity" work is done primarily by the creators, with some connectors involved so they can take the baton easily to the second stage. These people operate best in a loose or free environment. This allows the release of subconscious ideas that may have been stored for a long time. You should include a developer and a doer to keep your feet on the ground, and also so that they understand where these ideas came from when you get to the later stages of the process.

To be competitive in the marketplace, we need to know what effect the complex change around us is having on our customer. The tool for doing this is customer satisfaction monitoring, but it is rarely used correctly. Most ask their customer questions along the theme of "are we fulfilling requirements" because a customer agreement revolves around "requirements." The innovator wants to know the unfulfilled customer needs, and will instead ask questions such as "Where do you waste time?" "What do you have trouble getting done?" or "What are your biggest problems?" This opens a new opportunity and can easily be fitted into a customer satisfaction interview.

Technology is also giving us ways to get inside the customer's mind with instant messaging techniques such as Facebook and Twitter. Social networking increases diversity of ideas and increases market access.

"THE SCAN"

Classic strategic planning assesses the effect of change on the organization. The innovator assesses the effect on their customer. The lobbyist will attack changes, especially at a governmental level. The innovator will use the effect of these changes to identify new customer opportunities.

Change at a strategic level is affecting your customers, and it is essential to identify those changes before you talk to your customer. You need to know the legislation, the technology, the foreign competition that is affecting your customer. With this knowledge you will have an "informed" conversation. Your customer will respect your knowledge and be more willing to show their own "pain." The tool for identifying these macro effects is the *environmental scan*. The changes this scan will address are those caused by (1) the economy, (2) government, (3) law, (4) technology, (5) the environment, and (6) socio-cultural factors. Issues with suppliers and stakeholders should also be addressed.

You will find these opportunities by being "out there." Guy Laliberté's insight, which spawned Cirque du Soleil, came when he worked on the street with his buddy as a fire-eater! Exploration is a critical activity at the first stage in the process, but so also is interaction with others. You don't have a new customer yet. Your mission here is finding the opportunity, and this may emerge only after some searching. Note taking is the other critical activity at this opportunity stage. Look back at your notes on a periodic basis and you will be amazed at some of the "Aha!" moments that arise. I will talk more about this in Part III of the book.

There are many tools associated with the scan, but for the innovator one of the most important is that of *peripheral vision*, which asks questions such as:

1. What are the previous blind spots?
2. What other industries give us analogies?
3. What signals are you rationalizing away?
4. What are mavericks saying to you?
5. What are peripheral customers and competitors saying?
6. What surprises could really hurt you?
7. What new technologies are game changers?
8. Is there an unthinkable scenario?[3]

The scan identifies changes affecting key customers and shows where they will experience pain. The innovator addresses that pain. Plato said necessity is the mother of invention.

BIG DATA

With advances in Big Data, we can now gather and analyze enormous amounts of data. These data can be used to find new opportunities.

Rolls-Royce used sensor technology and data management to identify airplane engine problems at an early stage, optimize maintenance, and improve engine design. This enabled the company to develop a business model in which it retains ownership of the engines and provides maintenance, billing airlines a single fee based on hours flown. The data from the sensors also enabled better parts inventory management.

The UK is experiencing a major social problem as a result of online retailing. The "White Van Crowd" delivering packages from e-retailers to city residents is creating roadway congestion, and the drivers—on minimum wage with bonus—are extremely aggressive. The drivers themselves often have limited ability in English. The Greater London Council has set up one program that it hopes will inspire brand-new ways of doing business, launching the Agile Urban Logistics project. The project combines data on deliveries from the retailers with data on traffic conditions and optimization software. The goal is to encourage the private sector to develop new business models, such as shared-delivery services in specific areas.

Again in the UK, a partnership was formed between Vodafone, a phone company, and TomTom, a GPS provider. Vodafone can identify which of its phone users are driving, where they are, and how fast they're moving. The data can be used to pinpoint traffic jams, information TomTom buys from Vodafone to tell its own users where the jams are.[4]

WHAT WOULD MAKE YOUR CUSTOMER'S LIFE EASIER?

Customers are a rich source of knowledge when it comes to finding opportunities. However, you do not go to the customer and tell them you've come up with a really cool idea and wouldn't they love to have it! You don't ask what the new product is that they are looking for. The customer will think in today's context, not tomorrow's.

Instead, ask what would make their life easier. You need to be brainstorming with customers on questions like:

- What products or activities cause your biggest hassles?
- Which products or activities cause you to waste time?
- Which products or activities lead to problems being dropped on you?
- What are you having trouble getting done?

These are not easy questions to answer because customers will have learned to live with many of these problems. From the answers you get, you define exactly what the problem is.

If you are not yet feeling adventurous enough to ask these customer questions, you can still start the innovation process by conducting internally a cost of quality assessment, which is described more fully in the *Cost of Quality Implementation* webinar provided by ASQ Learning Offerings. The webinar described the "first cut" method in which you capture the collective knowledge of people inside your organization and identify processes where time is wasted, processes that cause hassles, or situations where problems are dropped on people. This again provides fuel for the innovation process.

THE PAIN STATEMENT

The output from this first stage of the innovation process is not some "fuzzy feeling"; it has to be a "pain statement." The problem must be given an initial definition. The definition will be refined in the next stage of the process, but for now we need that pain statement that captures the unhappiness, real or subconscious, of the present customer, whom you intend to move to your new market. This is not easy because the customer will have gotten used to the pain. Henry Ford's customers were used to feeding their horse every day of the week even though they only used the horse once a week. They had gotten used to paying vet bills. The horse manure was actually useful for growing crops.

The pain statement also needs to have some magnitude attached to it. I talk about measurement in Chapter 17, but for now, think of metrics in terms of *problem frequency* and *problem impact*. Generally, creators are not the best at problem definition, so the full definition should be left to the connectors. I describe this further in Chapter 7.

START WITH YOUR CORE CUSTOMERS

This builds relationships and provides the additional bonus that you keep out the competition. This develops your skills. You then move on to your other customers, especially the customers who are very demanding and you have difficulty supplying. They will "tell it like it is." Remember, finding pain is finding opportunities. This is where you find new needs. When you move on to people who are not yet your customers, you will find tomorrow's opportunities.

There are limits to the innovation that can be achieved by working with today's customer. Once your basic innovation process is working and you become more experienced, you then need to extend it to the *open market* concept where finding opportunities will then be much easier. These creation nets recognize that there is more knowledge outside the box, which I explain in Chapter 18.

Technology is also giving us ways to get inside the customer's mind with instant messaging techniques such as Facebook and Twitter. Social networking increases diversity of ideas and increases market access. These techniques are more applicable with consumer-based offerings, and we must be careful to phrase our questions carefully and ensure that we have a large body of data when analyzing responses.

What technology does point us to is the importance of open networking and the *wisdom of crowds*. Also called *crowdsourcing*, it has received criticism, but there is little doubt that it will identify a market situation. Bookmakers on a race track have known this for years. They set the odds based on what the crowd thinks, not on what they, the expert, thinks.

T-shirt website Threadless asked 1.5 million customers for designs, and their customers then voted for the preferred design. This is not new. In the Renaissance the textile industry in Tuscany operated like this. However, you must first learn how to learn, and you do this with your immediate customers! Open market innovation will also connect you to more potential opportunities and solutions. I will talk about connecting in the next chapter.

Before you go to the next stage in the process, score your organization on how well it performs on this stage of the innovation process (see Figure 6.1).

	Creating is finding the opportunity	Strongly agree	Agree	Disagree	Strongly disagree
1	Our people frequently come up with good ideas on their own				
2	We find out what problems our customers experience				
3	We interact with outside people to find new opportunities				
4	We explore outside the organization for market opportunities				
5	The work environment makes it easy to put forward new ideas				
		×4	×3	×2	×1

Figure 6.1   Innovation process assessment.

KEY POINTS

- Using the core competencies—and that includes knowledge—of your organization is fundamental to successful innovation.
- Using core competencies does not mean you have to continue to offer the same product or service.
- It is easier offering a new product than to change your customer base.
- Finding opportunities is the job of the creators, who operate in loose mode.
- Include connectors in this work so they can take the baton to the next stage.
- Brainstorming is one of the techniques for finding opportunities.
- The environmental scan is a widely used technique for finding opportunity.

Continued

Continued

- Ask customers questions like "What are your hassles?" "Where do you waste time?" and "What do you have trouble getting done?"
- Start with core customers, broaden to all customers, then move to "not yet" customers.
- Open market innovation moves us to "not yet" customers.
- We should pursue opportunities because they are commercial and not because they are unusual.

7
Connecting to the Solution

If I had an hour to save the world, I would spend 55 minutes defining the problem and five minutes finding the solution.

—Albert Einstein

In a past life, I spent a large part of my career in the Courtaulds corporation, which is now part of AkzoNobel. Although not a high-profile name globally, Courtaulds employed close to 100,000 people worldwide. In my early years, I was with them as a chemical engineer. In later years, because of my artistic ability, I was chief executive of their Christy home textile brand. This was a powerful brand and had an 80 percent recognition factor in the UK.

One of my tasks was to visit the annual trade exhibition in Frankfurt, which was called Heimtextil. I would spend three days walking the exhibition halls viewing our competitor's products, and over that time I would also gradually rediscover my ability to speak German. In the year I am thinking of, the market opportunity was in new designs for duvet covers, in particular for the juvenile market. I was looking for new design ideas through the stimulation of the exhibition.

After three days of inspection and analysis I was finally leaving the snow, slush, and rain of Frankfurt, and I was on my way home. I was walking through Frankfurt airport and my mind was still heavily focused on the products I had been viewing, but it was starting to relax. I was starting to think about home. My daughters were still quite young, and they would be getting ready for bed. Their beds would be covered in books as they read and settled down to sleep. If you have young kids, you will recognize the picture. The "customer need," at the risk of sounding clinical, was "places to put books."

A third factor then entered this melange. I found myself following a student wearing those big baggy dungarees. You know the type, made of denim and with lots of loose pockets. Suddenly, I had one of those "Aha!" moments that I know you have had at some time in your life.

I connected dungarees, an exhibition full of duvet covers, and my daughters at home in bed reading books. I suddenly thought, "Wouldn't it be a great idea to make a kids' duvet out of denim, with pockets on the side for them to put books in!"

Now, I am sure as you read this you are thinking, "So what?" However, you have had this kind of "Aha!" moment yourself. It's exciting, and you want to rush off and share your excitement with someone. Sometimes, you find solutions like this and you didn't even know you were looking for them! Well, I had to wait until Monday morning when I was back in the office to share my own excitement.

So far, this is a classic story of how innovation should and does occur. On Monday, that changes. I walk in the door. I am the chief executive, and everyone courteously asks how my week went. Of course I say I've had this wonderful new idea, and of course everyone says, "Yes, what a great idea." Again, you have probably been in this situation.

This is not the way to move forward with innovation. All too often, it is the boss's idea that moves forward. In truth, the collective knowledge of your people is far more powerful. Linus Pauling, who won two Nobel prizes, said his success was not due to better intellect but due to better networking. He was a passionate believer in collective knowledge.

The developers in our company worked on the idea and improved it from my original concept, which is good. The fabric was made lighter, and we used a technique called cross-dyeing to make it commercially viable. What I am describing here is the development stage of the product, which I will take you through in Chapter 9, and which I think we did very well. As a final point, I would add that the delivery stage, which I will discuss in Chapter 10, we did *not* do well. We actually went on to create and patent the duvet cover for kids that I had conceived. I am still proud of having that patent to my name. Nevertheless, I am sure there were other better ideas in the minds of our people.

DEFINING THE PROBLEM

In the previous chapter I talked about the pain statement, the frequency of occurrence, and the impact that are all part of the problem definition. Before we rush into burning energy on solutions, we must identify the size and nature of the opportunity.

Groupon saw an opportunity with the tight economic situation, and Guy Laliberté in the entertainment market. However, what is the size of the opportunity? How many people are potential customers, and how likely are they to adopt a new solution? Data matters.

Describing an opportunity when the customer does not yet exist isn't easy. The even bigger challenge is not rushing to potential solutions, which we are all prone to do. Defining a problem in data terms avoids this risk. Attaching a financial figure really gains attention. We also have a tendency to define solutions and not problems.

If you take your car in for repair, you might say there is a noise that sounds like a broken exhaust and the engine cuts out at stop lights. That defines the problem. To say "my engine needs a tune up" or "I need a new exhaust" is defining the solution. Define your customer's problem, and you are then ready to move on to the solution.

Learn about the customer problem through trial and error. Build strong relationships with customers. These are a major source of advantage. Focus on the customer experience. Customers want good experiences and complete solutions to their problems. Experiment and learn and be prepared to make a shift as new discoveries happen.

This is the preferred hierarchy of customer interface:

1. Data analysis
2. Talk to the customer
3. Observe the customer
4. Be the customer
5. Involve the customer in the design

The customer pain will express itself in terms of:

- Cost and time
- Effort and emotion
- Risk and worry
- Obstacles

In short, "What keeps them awake at night?"

CONNECTING

An interesting fact about problem solving is that there is no correlation between intelligence and problem solving. There is hope for us all!

Scott Page, a professor at Caltech, researched this back in the '90s. He put together a group of Mensa-level people and a group of "ordinary people," what he called the "Brown Socks" group. They were given a series of problems to solve, and the Mensa group was repeatedly beaten by the Brown Socks group. The Brown Socks group had *additive knowledge*; the Mensa group had *identical knowledge*, which was not additive.[1]

The connecting I will describe recognizes the universality of knowledge and that someone somewhere has probably already thought of the solution you are looking for, but in a different context. That is the power behind networking for collective knowledge.

This means connecting to other business environments where social trends your customers are experiencing may have already been picked up. Networking, which I describe in Chapter 14, is one important way of doing this, but the fundamental point is that breakthroughs occur at the intersection of bodies of knowledge—the "spark of ingenuity." This is about understanding the customer experience.

I mentioned previously some of the social shifts that can influence market need, and they came from connecting to other environments. Groupon saw an opportunity with the tight economic situation and went online with their coupons, Amazon exploited the growth of online retailing to provide nonperishable commodities, Guy Laliberté saw change in the entertainment market.

I suspect that Yellow Tail Shiraz connected to the social move toward drinking more fruit juices; I wonder when the airlines will see the social move toward drinking coffee in a pleasant environment. When will that short-haul air flight become like sitting in Starbucks? Some serve Starbucks coffee but miss the fact that *the experience matters*.

THE MYTH OF EPIPHANY

When we have insights, people think the solutions come from some kind of epiphany. Epiphanies are in fact the last piece of the jigsaw puzzle, and they arise from previously working on a problem. Archimedes's famous "Eureka!" came after working for a long time on how to establish the density of the gold in the King's crown. You can see why the loose mode is essential at this stage as well. Radical solutions come in all shapes and sizes.

If you look at history, most innovations build on previous experience, and most innovations also come as a result of stumbling through bad decisions. If you want breakthrough innovation and want to avoid this stumbling, you should determine what aspects of your present product or service block the solution. This is where you sacrifice what have been regarded in the past as sacred attributes of your offering. If you look at successful innovations in the recent past, all had the courage to do this. They all removed sacred attributes of their product.

In retailing, IKEA took out sales staff, Southwest removed first class in air travel, and Cirque du Soleil eliminated animals. However, they added something new after looking around and finding an unsatisfied customer need. IKEA added daycare and a café, Southwest added leather seating and entertainment, Yellow Tail Shiraz made their product fun, and Cirque du Soleil developed sophistication. These solutions don't seem radical today, but at the time somebody said, "You can't do that!" Remember, you are "sacrificing the sacred." This willingness to change requires the same loose or free environment of the creator stage.

This work on conceptual solutions is done primarily by the connectors, with some creators staying involved. You should include some developers so they can take the baton easily to the next stage. Again, include a doer to keep some sanity and also so that they understand how the idea developed in the first place. You can now start to see how the composition of the innovation team shifts as we move through each stage.

At this stage you must find and try alternative solutions. One is not enough. Remember, the best way to get a good idea is to get lots of ideas. You will stay with the concept and evaluate the concept. Call it prototyping if you like, but at a very conceptual level. You are not yet developing a working product or service. You will look at solutions that involve "moving context." Remember Henry Ford seeing that meat processing factory and getting the idea for automobile production. There are a number of tools for doing this.

We start problem solving typically by collecting data to understand the problem, and we process-map to get a picture of the processes concerned. The shortcoming is that we then analyze the data and we get solutions that are within the context of the status quo, a broken step in the process, or an inadequate understanding of requirements (see Figure 7.1). Instead, we should step out of the box and look for fresh ideas.

Albert Einstein was so insightful when he said, "No problem can be solved with the same level of consciousness that created it."

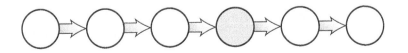

Figure 7.1   Failure in classic problem solving.

Traditional problem solving will tend to focus on immediate cause and effect, whereas the innovator recognizes the butterfly effect. I explained earlier that creativity calls for creative solutions dealing with two and three degrees of separation from the original causes. Connectors find solutions in completely new environments. Henry Ford saw meat hanging on hooks for transfer of material between work stations.

There is a popular myth that innovation comes from the lonely genius such as Thomas Edison. The truth is quite different. The idea for the screw fitting on a light bulb came from one of Edison's colleagues unscrewing a drink bottle. We have learned that the best way of "connecting" to these solutions is through collective knowledge.

IDEATION

Let me explain how to develop creativity and I will start by telling you that you are more creative than you think! Let me explain. Creative techniques called *ideation* or *idea creation* unlock the subconscious mind. There are a number of techniques available to harness collective knowledge, and the best known is brainstorming. A fashionable article was written in *New Yorker* magazine that stated that brainstorming does not work.[2] It does not work if the correct process is not followed. Over the years the process has been developed and flows thus.

During my "Introduction to Innovation" workshop I take people through an exercise that shows how you can be far more creative than you realize when you are in a group context. The exercise has a structure that owes a lot to the original ideas of brainstorming, which were developed in the 1970s and have been lost over time. There are three main inhibitors to good brainstorming:

1. A tendency to focus on one issue

2. Inhibition about providing revolutionary ideas for fear of ridicule

3. "Cruising"—a member of the group not contributing

The approach I am going to describe goes a long way to overcoming these problems.

THE SOCK EXERCISE

The group is asked to find alternative uses for a man's sock. The exercise takes about 10 minutes and it has five stages:

1. I first ask the group to write down three or four uses for a man's sock; each member does this on their own. Most people struggle to find four, the majority think of three. Examples might be to polish furniture, or to use as a woman's sock, an oil rag, or a cash holder.

2. I then ask each person to turn to the person on their right and share their ideas and build their list. You now hear laughter starting. The list grows: as a bag, for a glove, to tie something, to light a fire, to mop a table, or to gift wrap something.

3. I then ask the people to turn to the neighbor on their left-hand side and again share their ideas. By now, the buzz and laughter in the room are phenomenal. And the list grows again: as a Muppet, a weight, a ball, a slingshot, a doll's hat, a pillow.

4. I finally ask each person to turn back to the person on their right and build a final list. They add a toy and a mask. From originally struggling to find three ideas, every pair of people now has a minimum of 10 ideas and some have as many as 12 to 18 distinct ideas, allowing for the fact that many of the original three or four ideas will have been repeated on each person's list.

5. We then capture the ideas of everyone in the room. We will have 30 to 40 ideas, compared to the original list of about three per person.

This is collective knowledge at work!

Loosening up is a key component in creating what Edward de Bono calls *lateral thinking*. Icebreaker exercises are used for this and are well known. I like to use the "improv" approach that you find in the theater. Someone makes a statement like "It's a great day out there," and, importantly, this is not a question. The next person then says, "Yes . . . *and* . . . I think I will go for a walk." Then the third person might add, "Yes . . . *and* . . . I think I will wear my new shoes . . ." And so it goes on. Each statement builds on the last, and each statement is positive. It creates a great mood of creativity.

Research shows that the average adult thinks of three to six alternatives for any given problem. Get 12 to 20 people at a table to write down three or four ideas of their own, and then go through a series of steps until they get up to 40 *different* ideas from the group.

However, pushing beyond 40 to 100 ideas forces people beyond the obvious into "wild" territory. The key issue with all of this work is volume. As Linus Pauling said, "The best way to have a good idea is to have a lot of ideas."

The reason for this process is that it allows people to make crazy suggestions in private and avoid ridicule, but gradually share those ideas and build on them.

There are a number of methods, such as brainwriting, word association, and the Japanese NHK technique.

To be successful at ideation, we need:

1. *Knowledge.* Do people know the problem?

2. *Diversity.* Will solutions get challenged?

3. *Disruption.* Are disruptors present?

Ideally, we need at least 12 to 20 people in a group, and operating virtually we can go to hundreds. We also need to include customers and suppliers. It is like the old "think tank" technique of mixing disciplines to create that "spark of genius." We have learned that breakthroughs occur at the intersections of bodies of knowledge.

Excellent companies are recognizing the importance of allowing this "looseness," and while it might sound like a contradiction, looseness must be "structured" into the organization. This "group genius" has to be practiced regularly.

In the sock exercise I then ask each pair of people to select from step 4 one item from their list that is the most unusual, and one item that is the most commercial. Interestingly, these are rarely the same item. By the way, this is one of the common traps for innovators: we pursue ideas that are unusual and not ideas that are commercial.

I will now tell you more of the rights and wrongs of ideation. Keith Sawyer describes brainstorming well in his book *Group Genius.*[3]

To be successful, it is essential that team members listen to the ideas of others. The workshop exercise I described ensures that this happens. In large-group brainstorming, participants often tend to be busy trying to think of their own next idea, or they "cruise" between contributions. By listening well, we allow other people's ideas to trigger subconscious ideas and experiences from our own mind.

It is also important to recognize that ideation takes time, whether it is a piece of music or a piece of software. We hear stories of bands creating a piece of music in an hour, but in truth that is probably from the convergence of a set of ideas they had worked on for many months. Each person in the band makes a contribution, and it is the same in ideation.

You can also see that ideas need to be built "bottom up" to gain from collective knowledge. What we call a breakthrough or epiphany will only occur after time.

TRIZ

TRIZ was conceived in Russia in 1946 by Genrich Altshuller. The methodology was developed by analyzing commonalities in a large number of published patents. His ideas were so revolutionary that he was imprisoned by Stalin, and the KGB used sleep deprivation as a way to break down his resistance. It is a testament to his inventiveness that he tricked his prison guards by cutting out two pieces of paper in the shape of his eyes, drew in his pupils, and placed them over his eyelids before going to sleep, so giving the impression of being awake.

After the death of Stalin, he was released and went on to be responsible for many hundreds of patents. TRIZ has had a resurgence of interest in recent years as people have focused on the challenges of innovation.

TRIZ is pronounced *treez* and is a very "left brain engineering" method of problem solving. It can be said to be analytical and not creative. It has also been good at revealing trends such as the move from the mainframe to PC to laptop, and from the power company to fuel cell. In that regard I see it as drawing on the laws of thermodynamics and specifically the theory of entropy, which addresses the tendency toward increasing disorder in a system.

There is a lot of highly technical material out there that is attractive to people who are uncomfortable with the creative aspect of innovation. However, unless you are an avid statistical process control (SPC) fan, be careful. Rather like trying to use SPC outside the context of a developed quality management system, TRIZ will fail unless you use it in the context of an innovation management system, which was described earlier in this book in Chapter 4.

One of the better treatises on the subject is "Finding Your Innovation Sweet Spot" by Goldenberg and Horowitz.[4] They talk of many of the issues I have addressed already, how customers usually request only minor changes, and how we try to overcome this with brainstorming and then eliminating radical ideas. They talk of how it is important to find ideas outside the business market and introduce what is a subset of TRIZ called *systematic inventive thinking*. This is thinking inside the box, and is based on *product elements* and *five generic innovation patterns*. These five patterns are:

- *Subtraction*, where we remove components from the product, as in the Slimline DVD player.

- *Multiplication*, in which we alter an existing component, as in the move from the Gillette Sensor to the Gillette Mach 3 razor.

- *Division* is the reconfiguring of existing parts, as with the move from the "big box" to modular stereo components.
- *Task unification*, where we assign an additional task to an existing component, such as with the combining of the car window defroster and radio aerial.
- *Attribute dependency change* is a little more complex and addresses the relationship between the product and its environment; an example of this is the photochromic lens that darkens in sunlight.

TRIZ is for the left-brain analytical thinkers. It can also be said that this approach does not guarantee radical innovation.

SACRED ATTRIBUTES

As a connector, we are presented with a need and tasked with connecting to a solution. There are two key subtasks. First, determining what is blocking the solution, and second, looking at solutions in other contexts. Let's look at *solution blockers*.

I presented examples of companies who did this earlier in the chapter. They removed sacred attributes of existing offerings. A sacred attribute may have been needed 10 or 20 years ago when the offering was conceived, or it may have just slipped in unnoticed.

Airlines are a good example of where social change and cost reduction have contrived to make the experience of flying today a total misery. It used to be that everyone was served a fine meal with smart cutlery, but with the onset of the TV dinner, a cost-cutter saw the opportunity to save money. This worked for a while, but the airlines' customers rejected TV dinners a decade ago, and yet the airlines keep providing them. Serving a meal was a sacred attribute from the days when air flights were largely intercontinental.

The traveling misery was compounded by aircraft becoming larger (to achieve the consequent cost saving for the airline) but with the interior of the aircraft incapable of handling the logistics of the increased passenger load. The airlines further infuriated 80 percent of their passengers by cramping them into seating that was an ergonomic disgrace while giving 20 percent of their customers more room than was necessary. Airlines like British Airways even offer different attitudes and behaviors to the privileged 20 percent of their customers. This is based on the assumption that the 20 percent have influence and the 80 percent will not be heard.

It is small wonder that someone saw an opportunity. British Airways was attacked by Richard Branson with his Virgin Atlantic airline, and the many North American airlines were attacked by companies like Southwest and WestJet. The "new market" carriers stopped providing bad food and cramped, uncomfortable conditions, but did provide entertainment and friendly service.

What customers crave are good experiences and complete solutions to their problems. Don't offer websites that make people angry, and don't use call centers on the other side of the planet—selected by an accountant—who have no understanding of local circumstances. They upset your staff as well as your customer.

WALK BEFORE YOU RUN

Sow the seeds of your innovation process at a level where you know you can succeed. I don't want you to stay locked in process innovation, but it is a good place to start. Let me tell you about one of my clients, Jim Laforet, who talked about how he started the innovation process in his own company, Spectra Energy, at a "Breakfast of Innovators" that I run on a periodic basis. Jim managed the customer service operation at Spectra, which is a gas distribution company. As Jim said, it is hard to be innovative with a molecule of methane. Add to that the fact that regulators do not reward innovation. However, Jim had "innovation" on Spectra's performance assessment and in the mission statement, so he brought his 24 supervisors to a two-day workshop that I put on for them. The workshop was similar to the one I describe in Chapter 13, but with extra time to actually start work on opportunities.

The group produced 45 innovation opportunities, of which 22 were implemented. The changes varied from restructuring of meter routes to simple things like paper shredding, but he got his people thinking in "innovation mode." The fundamental in all of this is that customer satisfaction impacts revenue in a call center. Happy customers pay on time and complain less, whereas unhappy customers do the opposite.

What Jim's team learned in the workshop and in subsequent implementation was that people needed to know their innovation roles. Innovation was a state of mind, and you have to keep it fresh.

They also saw that they could apply the innovative state of mind to their processes, and their "early win" was the elimination of 2000 hours wasted handling high-cost bills in a process they changed through rapid innovation.

Having generated volume, we then need to identify which solutions are most likely to fulfill the customer need but at the same time be most difficult for competitors to copy. Still working as a group, we use old-fashioned storyboards, which still work fine for group critique, and we use sticky notes. However, we must collect data to support our choices.

EVERYONE ELSE BRINGS DATA

Deming said, "In God we trust, everyone else bring data." This is not easy for creative people, so include developers, who are "data people." Some of the primary areas in which we need data—and most of this data will not be crisp—are as follows:

1. The time needed to develop a working solution. At this stage we are only at concept.
2. The probability of being able to produce a working solution.
3. The cost of producing a working solution.
4. How radical is the new solution compared to the current solution?
5. How easily can this solution be copied by the competition?
6. Do we have the in-house competencies to produce the solution?
7. Who are the new suppliers we will need, and what is the risk attached to those suppliers?
8. What are the delivery chain choices and, again, the attendant risk?
9. What price will the market stand?
10. The likely ROI of the new product or service.

I'm going to stop at that point, and those of you who live in the R&D world will be familiar with the necessary data for developing a new-product portfolio.

Much of these data will have wide tolerances attached, but that does not excuse us from collecting the data. It can often be helpful to provide data collection support to the folks working on the proof of concept if data collection is not their strength. Another thing to beware of is the introduction of bias into the data collection. This can happen so easily and is often hard to detect. Somebody sees a wild idea and says, "Oh, I'll kill that right away," and yet that wild idea may be the perfect answer to the opportunity we have found. As you know, I am an engineer, and we tend to be good at data collection and analysis, but we are not good at accepting change, and we are externally risk averse.

My key point here is one you have probably heard before. Fail early! The innovation process has two phases: a *creative* phase and an *execution* phase. The budget for the creative phase is typically only 20% of the total cost of getting a new offering to market. Clearly, the more we can weed out the weak offerings in the creative phase, then the better the cost outcome.

You must constantly be testing new ideas, some of which may even be in conflict. The ideas will be selected based on trends and the core competencies of the business. You are trying out different options for the market. You find what works, stop the things that don't work, and resource the things that do. Don't wait for the competition to make your products obsolete—do it yourself. The knowledge gained from this testing is a primary input to strategic planning, which I discuss in the next chapter.

You can see that if you approach this second stage of the innovation process in the right way, you will generate lots of ideas. The challenge is then, "Which ideas do we run with?"

This takes us to the tipping point in the innovation process, where we select our solutions. Before you go on to the next stage in the process, score your organization on how well it performs on this stage of the innovation process (see Figure 7.2).

Connecting is finding the solution		Strongly agree	Agree	Disagree	Strongly disagree
1	We work with clients to find solutions				
2	We find markets where there is no competition				
3	We use a defined problem solving process for our opportunities				
4	We find solutions by going outside the organization				
5	We find the best solution				
		×4	×3	×2	×1

Figure 7.2 Innovation process assessment.

TEST YOUR IDEAS

Start with a broad idea about where the solution to an opportunity may be. You will follow stages of testing, information gathering, and changes in direction, or "pivoting," as the model is revised to arrive at the final, validated version. Typically, the conceptual solution changes radically as the testing unfolds.

> ### KEY POINTS
>
> - Using collective knowledge is the most powerful way of finding alternative solutions.
> - Most innovations are built on previous experience.
> - Be prepared to sacrifice sacred aspects of your current offering.
> - Somebody will be sure to say, "You can't do that."
> - An epiphany is really the last piece of the jigsaw puzzle.
> - Connectors find conceptual solutions, but it helps to include developers who can take a solution to the next stage.
> - TRIZ is a systematic approach to finding solutions and is preferred by left-brain thinkers.
> - Right-brain thinkers prefer ideation techniques.
> - Ideation, even though it is loose, follows a definite process.

8
The Tipping Point

I shall be telling this with a sigh
Somewhere ages and ages hence:
Two roads diverged in a wood, and I—
I took the one less traveled by,
And that has made all the difference.

—Robert Frost

I think back to my younger years when I used to play Monopoly. They say Monopoly brings out your dark side; I recall the high stress I used to feel over what was "only a game." I used to win a lot because the game contains some marginal decisions, and yet from time to time, no matter what I did, I couldn't help but lose. Maybe I got stuck "in jail" while everyone was picking up properties, or just kept landing on properties that someone else owned. You can not control the universe, but you can eliminate unnecessary risk by the use of data.

I have a strong aversion to gambling, and yet throughout my life I have been an explorer and risk taker, whether it has been being stranded in the African bush through car failure, walking through a "whiteout" in a wilderness through weather change, or escaping a major city riot with buildings burning. In my younger years I did not calculate risk too well. Now I do it more carefully. Innovation is where you take the marginal decisions and take calculated risks.

Risk taking is different from gambling. *Risk taking* is where you assess the likelihood of an event occurring and you assess its impact if it were to occur. If you want to innovate, you have to take risk. In order to assess risk, it will be essential to have data on potential solutions from the connectors. Risk taking does also have an intuitive component.

SELECTING SOLUTIONS

Malcolm Gladwell's book *The Tipping Point* is a great read, and its title introduced a major new term into the world of business. *The Tipping Point* shows how small occurrences have a major impact on subsequent events.[1]

You are taking major risks at this tipping point in the innovation process, and it is critical that the attached risk is both measured and then managed. Although, I will emphasize, *do not deny your intuition*.

Gladwell's *Tipping Point* is about how an idea, trend, or fashion crosses a threshold and spreads like wildfire. It's like a video on YouTube going viral because it catches everyone's attention. That's the point we are now at in our innovation process. The selection of which opportunities and solutions we will pursue is the "tipping point" in the innovation process. This is where you want to catch everyone's attention.

There are four common failings in making these selections:

1. We do not make a selection, and then we pursue too many options and so under-resource our options.

2. We select without good data, and make choices based on the wrong criteria. An idea may be unusual and gain attention but have no commercial merit (remember the sock exercise in the earlier chapter).

3. We select in a risk-averse mode and pick only short-term options where we are competing in a crowded marketplace.

4. We only assess risk internally and not externally (I will talk more about this in a moment).

Innovators risk their reputation when they develop "crazy" new solutions, and they risk ridicule from their colleagues and are often accused of being a "mad scientist." The media also make much of "accidental" breakthroughs such as Alexander Fleming's discovery of penicillin, and yet these "accidents" and "crazy" solutions are only a small part of the innovation process. The creative phase of the innovation process is where we find the opportunity and connect with potential solutions. This creative phase is, in truth, relatively low risk, and certainly low budget when compared with the execution phase of innovation where the risk increases significantly. Before we enter the execution phase, we must select our preferred solution so that in the execution phase we can focus on making it user-friendly before we deliver the solution to the market.

Risk may be short, medium, or long term. Xerox measured the period over which they achieve their required ROI on their innovations. Their median ROI on successful innovations was 7½ years. An accomplished innovator calculates risk, prioritizes risk, and mitigates risk.

If your offering is not user-friendly when it reaches the market, it will not succeed. Historically, this was the great strength of Kodak. George Eastman said, "You press the button, we'll do the rest." More recently, the first digital cameras were slow being accepted because they were not user-friendly. Daniel Kahneman was awarded a Nobel Prize for economics in 2002, and among the work he did, he showed that for a new product to be accepted it had to achieve a tenfold, yes, logarithmic, increase in perceived value to be accepted by the user. If you say you're offering is "twice as nice," you're not even close. What the late Steve Jobs did so well was that he understood the user. He created user interfaces that worked for real people.

The real risk for the innovator is not so much "will the product work?"—we deal with that in the development stage; the risk is far more "will the new offering be accepted?" and "will our new business partners deliver?"

Remember that statistics show that out of 3000 ideas, only one will make it. Unfortunately, we often kill the best ideas with overaggressive requirements for ROI. On the other hand, we often let a high-risk offering through because we failed to assess the risk attached to a potential new offering. We need to make the right decisions on where to invest our resources.

SELECTION AND STRATEGY

The strategic planning sessions of your organization are where you make the decision on what to select. We need enough data and information input to this session to be able to select based on risk and make decisions on which products to move forward. In the packaged goods industry, 30,000 products per year are introduced, and between 70 percent and 90 percent don't stay on the market more than 12 months. This is partly due to test marketing, but also due to a lack of sound risk assessment.

Many CEOs complain that their strategic planning process does not produce new ideas and is often just a game of politics. The annual strategy review is often a series of adjustments to last year's numbers with very few new ideas. It does not prepare for the risks ahead or serve as the focus for the company's direction. Strategic planning must actually make strategy and encourage risk taking.

There must also be quarterly reviews of strategy in which the mix of the portfolio is checked and the expected ROI and risk are also checked. This reduces the feeling of the annual review being a "once a year only" activity. This is done by your core group of strategic planners, who are usually rising stars in the business. I discuss this more in the next chapter on development.

SHORT-TERM THINKING

We need enough data and information input to the strategy session to do risk assessment and enable us to make forward-looking decisions.

Return on investment is good as a guideline but not to rank ideas for a decision. Just determine whether the revenue will be a hundred thousand or a hundred million. Think logarithmically.

The other thing to beware of at this decision-making point is short-term thinking. Most companies look for three years' ROI from any new product. The cell phone was launched in 1985; widespread payback did not come until 1998.

De Smet, Loch, and Schaninger in *Anatomy of a Healthy Corporation* talk of the pressure with which we are all familiar: the pressure for short-term performance and the tendency to react to what is "in your face" overriding the need to invest in long-term thinking. Their research shows something that intuitively we accept but in practice we all too often reject.[2] The portfolio of innovations in which we invest must contain a mix of short-term certainties and long-term risks (see Figure 8.1).

The low-risk projects of two years or less are very much driven by immediate customer need. These rarely qualify as breakthrough innovations. However, the strategists must also invest in wild-card opportunities in the short term, where the risk is high and learning will be fast. Financial institutions such as Capital One have become good at this. You must be prepared to fail.

Your portfolio must also include the longer-term five- to seven-year projects, both low- and high-risk. This is your "seed corn" for tomorrow's business. These don't all have to be high-risk. The long-term selection comes from viewing market trends.

My good friends at Canadian Standards Association decided several years ago to invest in developing a guidance document on global warming. Some people looked at this as a make-work project. The product was developed and completed, and then six months after completion, Al Gore's

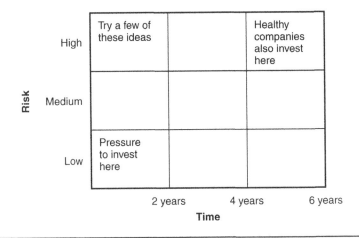

Figure 8.1   Project risk versus time scale.

An Inconvenient Truth hit the media, and CSA was in pole position at the racetrack. Some would say that was good luck. I would argue that this is looking at the significant trends in your industry and recognizing the core competencies of your own business. This is not just a job for the CEO; it is a cross-functional activity.

What are your own strategic influences? The economy, China, demographics, digitization, the environment? You need to be evaluating their long-term impact on your business. Your environmental scan, which I described in Chapter 6, should be extrapolated to show the medium- to long-term effect of the changes your customer is experiencing.

In setting strategy, look to your strengths. Volkswagen has a history of delighting customers and limiting environmental impact. As a twenty-first century automaker, its brand elements are innovation, responsibility, and value. Historically, VW innovation was considered the province solely of engineers in product development, and did not include people in market research. This is a common problem in engineering- and product-focused firms, but that has changed at VW.

They have "stretched" their brand by embracing Škoda for their volume car and Audi for their luxury brand. Volkswagen owes much to a strategy whereby its cars share components. This reduces variation in demand for individual components. It can switch production at its plants from one model to another whenever the demand for car models shifts. I visited the Škoda factory near Prague and saw how VW has transformed what was the laughingstock of Europe into a leading brand through this strategy. Add

Table 8.1 Innovation focus by industry sector.

	Improvement	Minor innovation	Radical innovation
Technology	45%	40%	15%
Diverse corporation	70 %	20%	10%
Consumer products	80%	18%	2%

to that the collaboration with the local university on research areas like robotics and water repellant surfaces.

Nagji and Tuff did some interesting assessments and offer these ratios for a new-product portfolio. They show what they see as an advisable split between *minor innovation* and *radical innovation*, and show different ratios for three different industry types: technology, diverse corporation, and consumer products (see Table 8.1).

They see the technology sector as clearly requiring a higher percentage of radical innovation in its new-product portfolio, while they see consumer products as requiring a relatively low percentage.[3] I would argue that companies who have taken a more aggressive stance in consumer products have beaten the competition. We also need to ask, is Apple considered *technology* or *consumer*?

BEWARE OF "MINOR" INNOVATIONS

Business in the '90s became obsessed with "minor" innovations and deluded itself that it was being innovative. This was all under the mantra of continuous improvement. Business has become obsessive with squeezing the last drop of juice out of the lemon of efficiency. Meanwhile, the world has moved on, and your market has changed. Research by George Day has shown that through this period the percentage of major innovation in portfolios dropped, typically from 20 percent to a little over 10 percent.[4] We became risk averse. On the other hand, the same research shows that in a similar period, 60 percent of the profit from innovation came from the major innovations. Clearly, we must aim for major innovations. This means moving beyond the "baby steps" I described earlier in this chapter and starting to be bold.

Major innovation means a product and/or a market that is new to the company. We just looked at risk versus time, and clearly risk increases

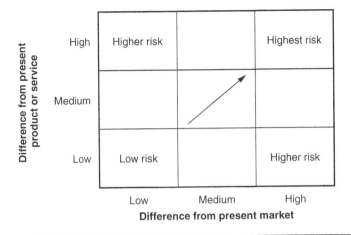

Figure 8.2 Project risk versus familiarity.

the farther out in time you are projecting simply because you predict future events with less and less accuracy the farther out in time you go.

Risk also increases as you become less familiar with the potential product or market (see Figure 8.2).

The higher the risk, the higher the chance of failure. Remember, however, that Intel says if they are not getting 10 failures to every success they are not taking enough risks.

In order to assess whether your product and market involve risk, you need to ask yourself the questions in the survey shown in Figure 8.3.

Your maximum score is 20 in each case. Low risk is less than seven points and high risk is more than 14. You will keep revisiting these questions as you conduct your quarterly review. This will show you how risk is shifting.

You also need to look at the potential ROI attached to the risk you are taking, and as I mentioned earlier, look at revenue in order of magnitude.

Clearly, your perfect world is low-risk/high-ROI, and the nearer you are to the bottom right corner of the chart in Figure 8.4, the more attractive the project. I would also argue that in the world of the innovator, the bottom left (low-risk/low-ROI) is a waste of time, and the real challenges come in the top right corner (high-risk/high-ROI) (see Figure 8.4).

The tough question in all of this is when, whether, and how to kill an idea. A sterile risk analysis is one way, and this is better than a simple ROI. The question is more *when* to kill it than *whether* to kill it. The tools I have just described as part of the strategic selection process need to be

For the product	Strongly agree 1	Agree 2	Disagree 3	Strongly disagree 4
We have the knowledge/technology				
We have the development capability				
We have the competencies				
We have intellectual property protection				
We are capable of production/delivery				

For the market	Strongly agree 1	Agree 2	Disagree 3	Strongly disagree 4
We have the business (supplier) partners				
We have the delivery mechanisms				
We have early adopters				
We know our competitors' positions				
We know the customer decision process				

Figure 8.3 Assessment of risk.

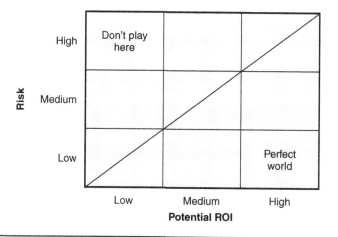

Figure 8.4 Project risk versus ROI.

fueled with data that were developed during both the opportunity and solution stages. The tools themselves should be used as assessment guidelines during the earlier problem-solving stage. The aim should be to pursue two or three primary options during the development stage, which I discuss in Chapter 9. During this stage the best option will emerge as more data and knowledge are uncovered, and the other options will be shelved.

INTERNAL AND EXTERNAL RISK

One other perspective on risk needs thought, and that is internal versus external risk. We usually assess internal risk well because we are close to it, but external risk has far more impact because we are less able to mitigate it. One thing that Edison and Bell both did well was take care of the infrastructure that their new ideas, the light bulb and the telephone, both needed.

Risk needs to be evaluated in the present selection stage and then addressed in the development stage, which is discussed in the next chapter.

The first external risk is with suppliers and subcontractors. Do you know who they will be? How reliable are they? Do you have backup? The chances are that if you have a new product or service, then you have at least one new supplier or subcontractor, and they may not be conventional. It may be software or knowledge that they are supplying. All your other suppliers can keep their promises, but if this one fails, you fail. Assess your supplier. If you have a QMS, you know how to do an audit. Don't just assess their product or service, evaluate their management system and their design and development processes. If you don't have the audit capability, get an ISO 9000 registrar to do it for you. If they have weaknesses, mitigate the risk.

If you have four key suppliers with three of them at 90 percent probability of delivering and the fourth at 40 percent probability, then the combined probability of your supply base coming together for you is only $0.9 \times 0.9 \times 0.9 \times 0.4 = 29\%$! You certainly wouldn't move forward with an internal risk at that level without serious mitigation.

At the other end of the business, you must assess how many hurdles you must jump before you reach the ultimate user. A classic modern story here is Michelin with its "run flat" tire. The *original equipment manufacturers* (OEMs) were not close to the development and so then needed another three years to include it in a new vehicle from their own design start. The OEMs were cautious and so introduced it in a limited number of vehicles. On top of this, garages needed special equipment and staff training, so by 2005, and after more than 10 years from the concept stage, the product had only limited use on a restricted number of vehicles. The good news for Michelin was that they did find a customer in the military. If you got a flat tire in Kandahar, you would much rather drive back to base on a run flat tire than stop and change a wheel!

Who is your distributor? What is the decision time cycle? These numbers impact the ultimate time to market. If you are entering a new market, these cycle times will need even more analysis than your supplier capability assessment. These data are going to be far more difficult to obtain, and of course as you start asking questions it gives a heads-up to your customer and also to your competition. Maybe you need a business partner with market knowledge in your intended new market.

The lesson? You may well find it better to pursue a new product or service with high internal risk that you can manage and mitigate a low external risk that you don't have to manage.

INFRASTRUCTURE

Delivery chain risk then brings us to the question of infrastructure for the new offering. When Edison developed the light bulb, he still needed a power grid to enable it to be used. When Bell developed the telephone, he still needed cables for messages to be transmitted, but had the advantage that "the wire" already existed. One of the most radical innovations of the last half century was the British Mini. When Alec Issigonis recognized the market need for a motor vehicle that was affordable in a nation that was bankrupt after WWII, he simplified the vehicle by eliminating the transmission and creating the transverse engine, which had the additional benefit of engine weight on the drive wheels. The vehicle was priced at £499, or less than $1000 on today's currency exchange. Issigonis did not need to create infrastructure as the roads already existed, but special attention had to be given to the suspension on such a small vehicle. That was also a focus in making the very basic vehicle user-friendly. He adapted to an existing infrastructure. More recently, the success of the iPhone did not occur until after the infrastructure of iTunes had been created. Risk assessment must consider carefully the infrastructure within which the new offering will operate.

RISK MITIGATION

I don't need to say too much about this; it has become a well-established practice. To quote a few examples, you may find a single-source supplier who is slow delivering, so you manage this by holding inventory. You may be unsure about key competencies, and you invest in additional competency development. On the other hand, do not invest in soft budgeting. Just about every risk mitigation costs money. Experience shows that tight budgeting drives resourcefulness. The one resource that is irreplaceable is time, so plan carefully and thoroughly.

RISK AND SOCIAL RESPONSIBILITY

Innovation can provide solutions to large numbers of problems, but it can also be the cause of those problems. The financial sector is a good example of where both sides of the social responsibility coin are very evident.

The subprime mortgage problem is an example of a highly innovative financial product that was perfectly legal but which caused enormous social damage. Many are now pointing to the removal of financial regulations in the '80s and '90s as having allowed "subprime" to happen. On the other hand, one of the most radical business model innovations in the financial sector is the one we have seen in India. This involved setting up a completely new service structure reducing the access barriers for women. The Mann Deshi Mahila Sahakari Bank (MDMSB) is a unique cooperative bank run by and for women. A unique aspect of this fully computerized bank is that it offers weekly and fortnightly credit and savings plans to its customers, most of whom are daily or weekly wage earners. Unlike any other bank, it also provides daily loans for buying vegetables or fruits. This is an exciting innovation.

We now regulate pharmaceuticals very closely after the Salk vaccine horror of the 1950s. We regulate toys and vehicles for safety although it has been a hard road getting there. Why not regulate financials?

The core problem here is that the financial innovator is often driven by immediate personal greed rather than user benefit. For now, the solution is regulation, but prevention is better than cure. The long-term solution is to educate in socially responsible behavior through our schooling system. There will always be clever criminals, but the fewer there are, the easier it will become to police new innovations.

INFLUENCE CHART

In responding to change, you will seek change, and you must address resistance to change, especially within your own organization. We use the influence chart (Figure 8.5) to identify who will resist change and why. It is then the job of the president, who may be all in favor, but they may have a VP with

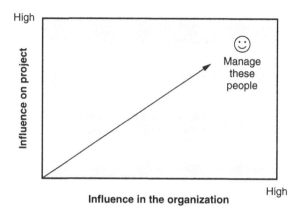

Figure 8.5 Influence chart.

a vested interest in status quo. You don't want to go the way of Kodak. This means first identifying what the resistor will lose and to address that loss.

Having made your decisions, the game now changes.

SWITCHING FROM LOOSE TO TIGHT

Once decisions have been made, we switch from loose to tight. The tipping point is where you actually change your mode of operation. So far, this has been a loose and flexible process, but now we've made some decisions, so we go tight. In the first two stages you will find lots of opportunities. Now you narrow your focus. A key task for leadership is monitoring this change in mode. We now need to move with speed.

Again, before you go on to the next stage in the process, score your organization on how well it performs on this stage of the innovation process (see Figure 8.6).

KEY POINTS

- Risk taking is fundamental to innovation.
- Risk can be measured, but there is an intuitive component.
- The tipping point in the innovation process is where we select which options to pursue.

Continued

Continued

- Organizations frequently try to pursue too many options.
- Choices are often made based on poor data.
- Organizations have become risk averse and only pick short-term options.
- External risk is often assessed poorly.
- A new-product portfolio must include a number of long-term options.
- There is a danger of being drawn into "safe" and minor innovations.
- A major innovation is an offering that is entirely new to the market.
- External risk is more difficult to manage than internal risk.
- Risk exists both with partners (supply chain) and customers (delivery chain).
- This is the point where the process switches from loose to tight.
- Risk is reassessed as you move through the development stage.

Strategic planning is the tipping point		Strongly agree	Agree	Disagree	Strongly disagree
1	We do not go for short-term ROI				
2	We assess external risk				
3	We assess internal risk				
4	We have sufficient data to make decisions				
5	We provide resources to support our decisions				
		×4	×3	×2	×1

Figure 8.6 Innovation process assessment.

9

Developing the Solution

Genius is one percent inspiration and ninety-nine percent perspiration.

—Thomas Edison

Above everything, don't fool yourself into thinking that because you have found an opportunity and also found a solution that there isn't someone else on the planet who has found the same opportunity and probably a similar solution. To quote the lawyers, "time is of the essence."

Alexander Graham Bell is probably the classic story of "the need for speed." He worked for many years with his colleague Thomas Watson developing a way of being able to send "sound over the wire" and being able to "talk with electricity." He was also well aware of patent requirements for intellectual property through his father-in-law to be, Gardiner Hubbard, who ran the Boston patent office. It is one of those legends of business history that he beat his archrival, Elisha Gray, to patent his device by only a matter of hours.

In most businesses, the market opportunities and the conceptual solutions are actually fun to find. It is the development process that is the tough, grinding work. Where most R&D operations go wrong is that they are disconnected from the real world. A lot of serious questions are being asked about the effectiveness of R&D. IBM's global survey has shown that little more than 10 percent of new innovation is coming out of R&D facilities. A lot of this is due to their isolation and obsession with secrecy. It is essential at the development stage to be connected with both customers and suppliers, as well as your own operations and sales people. Let me quickly revisit the early stages of the innovation process in the context of the business

I used to run, where there was a strong connection between the development and commercial processes. I would add that the threat of "Robbery and Denial," the competitor's idea of R&D, was just as great as in other industries.

In home textiles the market barometer was the clothing industry, and when I was in the clothing industry the barometer was always the couture collections that would push the envelope in color, garment structure, and fabric pattern. Many trends in fashion are cyclical and also relate to social mood and the economic cycle. At Christy we would source our own graphic ideas from artist colonies such as the one beside Lake Magiore in Italy. Let's say, for example, we saw an opportunity in paisley prints. We would look for alternative treatments by different artists. Remember, "The best way to have a good idea is to have a lot of ideas."

Having gone through stage 1 (opportunity) and stage 2 (solutions), we would then select. There is an overlap between selection and development as there is with the other stages in the innovation process. At the selection stage there may be a need to test alternative solutions, and you may choose to carry forward more than one solution. You may pick two alternatives for development, and very rarely pick three. This is where many people go wrong and take too many choices into the development stage.

My chief designer was Lorraine Robert from Montreal—a colorful lady, but tough as nails. She had arrived in England via New York and France. She was a brilliant artist, and I balanced her creative ability with a sense of market need. She also worked incredibly well with customers. Finally, I also brought the skill of color into the equation. They were exciting times. We would return from Magiore with our plunder and lay out the artwork for the upcoming development work.

Having selected the one or maybe two paisleys we wanted, we would run with the first step in the development process, which was to transpose them into a form that was (1) user-friendly and (2) could be manufactured easily.

You need to ask what these two things mean for you. User-friendly for us meant it would coordinate with color schemes being adopted by the consumer but at the same time give them a new and exciting experience. Ease of manufacture was the other critical consideration for the developer. Each color we added to the design increased cost, but one color is boring, two becomes interesting, and three starts to get exciting. A five-color print is extremely expensive, and also fraught with the risk of complexity. The magic solution is of course one color that is exciting.

In your own world, ask whether you are falling into the trap of too much complexity in your design and whether the service or product you are developing will be easy for your customer to use.

The irony of this development stage in the innovation process is that it is usually the most defined stage for most organizations and yet it is where most innovations get killed. One of two things happens: either the development stage is so slow that someone beats you to the market, or there is a tendency to revert to status quo and kill the new idea.

The development stage receives its input from the connectors after the opportunities have been screened by the strategists. Companies pour monies into development labs instead of into the creating and connecting stages, which are actually more critical and typically are only 20 percent of budget. A common saying for the innovator is "fail early." Development is an area where many companies are well resourced but are simply producing more of the same. A key to success here is having a well-designed development process. Whether you are designing a chemical plant or creating a fashion item, the stages are similar. I am going to take you through the stages for a new graphic design in textiles. You will relate to it as you are a user of the product and it is a fairly simple process.

DEVELOPMENT AND QUEUING THEORY

We try to fully employ our development people, who are a high cost, believing that it is efficient and that a busy development function will be faster and more efficient than one that is not using all of its people. Queuing theory shows that in highly variable processes like design and development, the amount of time projects spend on hold rises steeply as use of resources increases. This problem is because companies think the process is similar to manufacturing and order input, which are repetitive processes that also don't change much. The work-in-process inventory of product development is invisible, and most R&D inventory is shown at zero value on the balance sheet. As a result, business leaders do not have a good indication of "work done" and so often accuse R&D people of having time on their hands. The consequence is often pushing for more work to be done and a resulting project logjam in R&D.

To combat this problem, 3M has in the past scheduled product developers at 85% of their capacity. Google, when it was famous for its "20% time," enabled capacity to be available if a project fell behind schedule. There is a lot of pressure from the operations people to not do this. They are busy and resent the "space" that designers have. I remember this well when I was in my corporation's R&D division.

In practice, this means limiting the number of projects and making the work-in-process inventory easier to see with visual boards. It also means being ready to change the plan as new knowledge emerges from testing competing solutions. This is typical in innovation projects: customers' choices can also shift during a development project as they gain new knowledge.

THE DEVELOPMENT STEPS

Whatever the service or product your organization excels at providing, you have certain distinct steps in designing that offering. Those steps must be clearly specified, along with the "entry and exit" or "input and output" criteria for each step in the process.

Let me take you through the steps of developing a new print design. While you are reading this book you are probably surrounded by prints, whether in the fabrics in an airport lounge, the magazines in a coffee shop, the furnishings in your own home, or the clothes you are wearing. Look at one of these prints and let me tell you how they were created:

1. The original concept artwork will have been created manually or by computer graphics. Even with computer graphics, the artwork will normally need to be printed as a hard copy to show the color of the ink in true form.

2. The design will be extended to evaluate how it looks "in repeat." A graphic that looks good in a 12" square may not work when it repeats over an area six foot by six foot (36 square feet). The repeatability of the design is addressed.

3. Color normally comes next. The concept will be analyzed by the design team to select alternative colorways. The colorways will not be final, as new colorways will emerge during prototyping. Color balance is critical, and the ability of new colors to work with existing domestic colors is also vital.

4. The fabric strike-offs are next. How will the design behave on the final fabric to which it will be applied? Lighting effects like *metamerism* must be addressed; a particular pigment undergoes a color shift between daylight and artificial light.

5. Once satisfied with the strike-off, a short production run will be required. Depending on the printing technology, this may require investment in components for the printing equipment. This can be a significant scale-up cost. This enables identification

of production problems and also enables sufficient fabric to be produced for a short production run of garments.

6. The short run of garments is tested on a cross-section of customers, who will have been involved at earlier stages but not in detail. They may not have the visualization skill to see strike-offs in the final garment.

Clearly, you want as few iterations of this cycle as possible, and you want it to be as fast as possible. Recording design changes as you go through this process is something developers have learned from bitter experience. What developers do not do well is conduct a detailed design review to ascertain whether requirements have been met at the end of each development stage, and also concisely record the key issues at the end of each development stage.

I have deliberately taken you through a simple example so that I can explain concept and approach. You may be saying, "What has this got to do with designing a chemical plant or a new service delivery method?"

Look at the steps in the process and apply them in your own environment. It is essentially (1) develop concept, (2) scale-up, (3) test, on a repeating basis. Recording decisions and actions at each stage is vital to avoid reinventing the wheel in your development stage.

We may sometimes generate data by "modeling" or "piloting" a solution. Software guru George Box said, "All models are wrong, but some are useful." We must never lose sight of that, but still do everything we can to develop data that will guide us in our final choice of solutions. In my early years as a chemical engineer I worked on some exciting pilot plant projects. They really are great fun. I worked on desalination by vapor recompression, a radical form of countercurrent liquid extraction, and the early days of carbon fiber. We have to be careful to ensure that these projects do not take on a life of their own. They are there to generate data. At the same time, we must not isolate the people running a pilot. They must be connected to the bigger project.

We are at what is often called the "nightmare" stage of innovation. The development work has often been delayed through inadequate proof of concept at the conceptual solution stage, and we try to catch up the schedule by jumping vital steps at the prototyping stage. We even say, worst possible case, that we won't even pilot and delude ourselves—"of course it will work." Fatal error!

Fortunately, help is on the way for some people, especially in the manufacturing sector. It's called 3D printing.

We have traditionally made components for manufacturing using the techniques of the sculptor: drilling, milling, and grinding to work a piece of metal. 3D uses the techniques of the potter, building from clay up to the finished article. That is why it is often called *additive manufacturing*. A batch size of one costs the same as 100 or 1000.

This has huge benefits at the development stage of innovation. The eternal nightmare of the scale-up from development to production is now eliminated. It also enables the development folks to trial alternatives at low costs. An example is Timberland developing a new shoe sole in 90 minutes for $35 instead of spending a week and $1200.[1]

The technology combines the technique of inkjet printing, CAD software, and advanced materials. You've probably seen it in the movies. In *Mission: Impossible III* Ethan Hunt (Tom Cruise) and his team build a mask he will wear so he can impersonate the evil villain. You see the mask being built in a series of layers using a computer image and injecting a material onto a base where the material solidifies. They replicate the computer image on the mask.

In conceptual terms, that's all it is. In practical terms, we need the CAD software, which has developed well in recent years, we need hardware not unlike your desktop printer, and we need material.

The first step in all 3D printing processes is for software to identify cross-sections through the item to be built and calculate how each layer needs to be created. The machine then builds up the item, a layer at a time, by applying a thin layer of resin or by using a laser to melt metal powder. The material then hardens in the pattern of the cross-section, just like in the *Mission: Impossible* example. The tray on which the item sits then drops about a millimeter or less, and a new layer of resin is applied and the process repeats. A variation on this is like inkjet printing, the printing heads in the machine applying a liquid binder onto a bed of powder as color is applied at the same time.

This has many benefits: Cycle time reduction is the first. You make an item at the point where it will be used. You don't need a mold, which is the high-cost item in plastic fabrication. Material efficiency is much higher; you only use what you need. Most dramatically, economies of scale evaporate, and mass customization becomes a reality.

THE CHANGE FROM LOOSE TO TIGHT

This stage in the innovation process requires discipline. It is often a challenge to shift from the fun and freedom of the early creative stages to the

Chapter 9: Developing the Solution 103

data recording and analysis of the development work. For me, it was a shift from the freewheeling and wining and dining of Lake Magiore into the grind of dimensioning, cost analysis, and meeting deadlines. This is a vital shift you have to make.

This stage should be fast and not secretive, and should involve the ultimate user. Remember, this is where you move from loose to tight. Good project monitoring and control is vital.

As new knowledge is uncovered in the development stage, risk will change. ROI will change, time to market will change, and resource needs of time, money, and people will change. The leadership team must stay close to the development process and be personally engaged in the development review meetings. The meetings should be two or three monthly. The metrics that were the basis for the decision to proceed back at the tipping point must be closely scrutinized. If it is clear they are going the wrong way, we must not be afraid to kill a project. This is the most difficult of all innovation decisions and must involve business leaders if the decision is taken after the tipping point. The perfect "risk world" that we seek is shown in the three-dimesional risk diagram (see Figure 9.1). We want a highly radical offering with high ROI and low risk. We will rarely achieve this, but that is what we aim for.

The job for the developer is to make the solution work and make it user-friendly.

This development stage also requires the building of partnerships with your customers and distributors to enable the later operation of the delivery

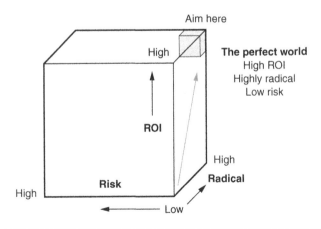

Figure 9.1 Three-dimensional risk diagram.

chain. This partner management must use data from the risk assessment conducted at the tipping point in order to manage relationships where the risk is highest.

This work is done primarily by the developers, with some connectors and creators staying involved to ensure integrity. You need to be including more doers by now so that they can pick up the baton easily in the last stage. However, you will see a tension building in the innovation team. You will recall how developers think "creators are not focused" and creators think "developers don't see the big picture." Keeping a team mix is important and leads to better results.

BEHAVIOR CHANGE

One of the biggest obstacles to acceptance of new ideas is behavior change on the part of the user. Research has shown that sellers overestimate the perceived benefit of a new product by a factor of three, while buyers overestimate the perceived benefit of an existing product by three times. The barrier to the new product ($3 \times 3 = 9$) means that to overcome this barrier we need a 10× increase in perceived value—an exponential increase.

Most new ideas struggle to find daylight because potential users resist anything new. The challenge for the developer is to make the new product or service so user-friendly that resistance is futile. The job for the developer is to make the solution work. Let me explain some of the challenges.

The psychology of behavior change is worth understanding. The classic approach in business to overcoming resistance to change is to create an attractive vision of the future. The word "resistance" implies that we have to "push." We will talk about creating the vision, or *value proposition*, in the next chapter, but first we need to put ourselves in the mind of the potential customer.

The book *Managing Transitions* by William Bridges brings out the issue very well.[2] In creating so called "resistance," people are actually saying there is something about the status quo that makes them very happy. We may have found what we regard as the coolest solution on the planet, but people will not go through unnecessary "pain" to adopt our solution. Back in June 2001 *PC World* magazine reported that over 60 percent of MS Office users were still using Office 97 and even Office 95, and Microsoft had taken the drastic step of no longer providing free support for these versions in order to pressure people to upgrade.

The developer's job is to address the pain of change and make the new product user-friendly. It is not to tinker and build in bells and whistles that the developer personally finds attractive.

RESISTANCE TO CHANGE

Daniel Kahneman won a Nobel Prize in economics in 2002 for his study of why people ignore what appear to be obvious benefits, whether in time or money. People actually look at a new innovation from the following points of view:

1. Perceived value—what they think it is worth.
2. From a reference point, comparing to something they know.
3. What gains and losses they will experience.
4. The opinion that losses matter more than gains!

This last point that losses matter more than gains is most important of all.

The research by Kahneman showed that people buying a product would typically have the mind-set of "just hang on to what you've got."[3]

An example of this resistance to change is the Dvorak keyboard. It increases typing speed by 30 percent, but who wants to relearn how to type? The ketchup bottle with its opening at the base seems like a great idea. The manufacturers even apply the label upside down to show us which way to stand the bottle. Wait for this . . . in the first five years of its life 70 percent of people still stood the ketchup bottle with the opening at the top and the label upside down! People do not change easily (see Figure 9.2).

If you want an example of an innovation success, look at Google. Who wanted a new search engine in the year 2000 with so many failures around?

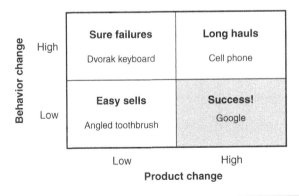

Figure 9.2   Behavior change and innovation.
Source: John Gourville, "Eager Sellers—Stony Buyers," *Harvard Business Review* (June 2006).

Google's algorithms are amazing, and hence its success. The user interface is so simple, some people no longer bookmark website addresses. No behavior change required! Google's meteoric growth depended in large part on its ability to innovate and scale up its infrastructure at an unprecedented pace.

We can all recall examples where our own resistance to change slowed down adoption of something.

I recall my wife buying me one of the first digital cameras from Sony. It was so difficult to use I continued to use my old roll film camera for a couple of years. This actually saved Eastman Kodak for a period of time. They hit the market with a camera that embraced the simplicity of the old box Brownie. It was so easy to use, and I recall my friend Trevor Smith, Global Quality Manager for Kodak, demonstrating to me the simplicity of their product.

Don't ever allow yourself to go to market with a product that is too complex to use. Simplicity and ease of adoption are your driving forces here.

Another way of looking at this is to aim for exponential improvement or benefit, a factor of 10. Think back to that research of Kahneman's.

We want the most radical new offering with the best ROI at the lowest risk. Call it utopia if you will, but that is where we aim. I represent this in a three-dimensional risk diagram (see Figure 9.1).

If you are aiming for "green field," then you are creating a new market, and you might think you have no resistance, but you need to think carefully. The iPod opened the market most easily with people who had not yet bought a car or a home stereo, the snowboard with people who had not yet skied, and the mountain bike with people who did not own bicycles. But this is more for the next stage in the cycle when we address the value proposition of innovation.

Before you go on to the next stage in the process, score your organization on how well it performs on this stage of the innovation process (see Figure 9.3)

KEY POINTS

- At the development stage, time is of the essence.
- The odds are that someone else has found a solution that will compete with your own.
- Development is overfunded and under-delivers in far too many organizations.

Continued

Continued

- A well-defined development process is fundamental to success.
- The input and output criteria at each step of development must be clearly specified.
- The essentials of development are (1) concept, (2) scale-up, and (3) test.
- Organizations often have difficulty moving from loose to tight mode.
- Partnerships with the supply chain and delivery chain are developed here.
- The primary challenge of the developer is to make the new product or service easy to use.
- Potential customers will significantly overvalue their existing product and undervalue the new offering.
- You will generally need an exponential improvement in time or cost savings for rapid acceptance.

	Developing is making the solution work	Strongly agree	Agree	Disagree	Strongly disagree
1	We fund novel projects				
2	We are willing to take risks with new ideas				
3	We are able to get people who can work on new projects				
4	We closely monitor project progress				
5	Our projects finish on time				
		×4	×3	×2	×1

Figure 9.3 Innovation process assessment.

10
Delivery

Time is of the essence.

—An honest lawyer

Over a century ago, people in Europe and North America used what we today call *huckaback* towels, which were made of a flat cloth with a honeycomb weave. The Christy family was a typical wealthy Victorian family, with each of the sons having a destined profession and any daughters being pointed toward other wealthy families for an "appropriate match."

Henry Christy was one of the sons of William Christy. He was sent on what was a statutory journey of the time. It was called the "grand tour." It was a journey across Europe, and its aim was to broaden the mind. Henry attained the outer limit of his tour when he reached Turkey and through his connections was able to meet the sultan.

Henry was given a tour of the royal palace and that included the "ladies quarters" or as we call it today the *harem*. He saw the ladies hand-weaving a loop-pile fabric that looked like a lamb's fleece when it was finished. It was used in the Turkish baths, for which that country was famous. Henry asked for a sample to add to the collection of memorabilia from his tour.

An idea was born.

Back in England, the Victorian population had just discovered the art of bathing, whether in a bath at home or in the sea while taking their holidays away from work. Henry's father William, back in the Droylsden district of Manchester, England, had a textile factory, and his brother was an accomplished engineer. Why not take this fabric back to England and encourage his brother to develop a weaving machine, or loom, that could make this fabric? Why not go into mass production and sell these very absorbent pieces of fabric to the Victorian population?

Henry's brother worked on the project and developed a weaving machine that carried two "beams" of yarn. The top beam moved faster than the lower beam and so created the loop pile, which in turn created the absorbency. This was the world's first towel, as we now know it, that was not hand-woven.

In 1851 the Great Exhibition was held at the Crystal Palace in London, and Christy's decided to exhibit their new towel. This exhibition was a celebration of all the great British innovations of the time.

Queen Victoria attended the exhibition, and as she walked through saw the towels. She ordered six dozen for the Royal household, and so was born the "Royal Turkish Towel."

This story has all the classic steps of innovation:

1. The exploration and the opportunity (The grand tour and Victorian bathing)

2. The connection to the solution (Harem and the Turkish bath)

3. Engineering and development (The Droylsden factory)

4. Delivery (The Great Exhibition and the "Royal Turkish" brand)

See how important the exploration was. Also, notice that although the element of "Victorian bathing" came late in the story, it would have been sitting in Henry's subconscious throughout the grand tour. He lived in a different world from his brother, but the two of them were "networked."

Notice the connection of two totally different environments, the exotic harem and depressing Droylsden. Networking between different environments is fundamental to innovation.

Notice also the "product launch," the purchase by Queen Victoria, and the creation of a brand. Where is *your* Great Exhibition?

The final challenge of the innovator is delivery, and be aware that the best ideas don't always make it! You now need a reliable delivery process as well as a "wow" product. Again, out of every 3000 ideas, only one makes it! Delivery needs discipline, and the organization must now become really tight. The delivery work is done primarily by the doers, with some developers, connectors, and creators staying involved to ensure integrity and provide advice and support.

This is where the operations and sales people take ownership, but remember, they must be involved in the earlier stages so that there are no surprises here. Production problems are eliminated during development. The value proposition that the sales people will use is also initiated during development!

In the Christy business that I just described and where I became chief executive, my works manager would be challenged with new manufacturing techniques or new outsourcing needs early in the development phase. The sales manager would have been engaging our close customers at the earlier connecting phase. He was always careful to engage those customers who we knew to be proven early adopters and never engage those customers who would plagiarize our new product with the competition.

The new manufacturing techniques might require new equipment and new operator skills. If you are in a service environment, this might mean new software skills.

You will notice that at this stage in the innovation cycle I refer to sales and production as a single entity. Infighting and finger-pointing have no part at this stage. These functions should be integrated. Their engagement in the earlier stages of the cycle and especially in the developmental phase is crucial for speed of delivery at this final stage.

When you have an organization in which marketing, design, production, and sales all connect well, then you have an integrated organization that is honest and effective. This is vital at each stage in the innovation process, and not least at this last stage.

We all have our challenges, and mine was my chairman. I believed in having a delivery system that our sales people could rely on; he believed that if you constantly "wow" the customer, they will forget your previous failures to deliver.

Delivery is payback time. If you have invested in a management system that has integrity, it will give you good data and information. If you have good information flows between the different areas of your business, you will beat the competition. Do not doubt that there is probably someone, maybe on the other side of the planet, who has thought up a solution to the opportunity you identified. Their solution may not be as good as yours, but if you don't move fast, they will beat you. They will even copy your superior product once it's out in the market.

At this last stage, again, time is of the essence. You will also notice that the loop has closed back to the sales people and also to the marketing people, who should be constantly seeking new opportunities.

At this last stage, you should know already who will be your early adopters. These are your first customers. You found them back when you were looking for opportunities. When it comes to selling down the delivery chain, you should have addressed the obstacles at the tipping point when you did the risk assessment. Too often, it is assumed that selling down the chain is just another selling job. This is a new product; if you have not prepped well, it could be a major failure at major cost.

The challenge for the sales people is to develop a value proposition and also collateral for the product. Supporting literature needs to be in hard copy as well as electronic format.

VALUE PROPOSITION

To place your product successfully with the customer, you need to specify your *customer value proposition*. This has become one of the most widely used business terms. However, many value propositions claim benefit to the customer but lack evidence. You need to know where you perform the same as the competition, where you are better or worse than the competition, and where you and the customer may disagree. You then concentrate on the two or three positive points of difference.[1]

Ultimately, the customer will adopt the new idea if they see that the new idea makes their life easier or more enjoyable. These are intangible benefits, and the key word is "benefit." We have spent so long working on features of the product through both problem solving and development that we can forget that the customer is interested in benefits and not features. Almost every basic sales course will tell a salesperson the difference between features and benefits.

The customer using a search engine is interested in finding information quickly, not in the wonderful algorithms of Google. A customer is interested in paying less money when they fill their gas tank and far less on the wonderful fuel-efficient engine. The value proposition has to focus on the customer's pain.

The value proposition has to focus on the pain and the customer's feelings. It needs to focus on two or three key points that will lodge in the customer's mind, and one point of similarity with existing solutions. This way, the customer connects to the existing solution and its shortcomings. They then read on to two or at the most three key benefits of the new solution.

The temptation is to list all benefits and create a shopping list as a catch-all for all customers. This approach has the major shortcoming that the list contains benefits that have marginal value to the customer and will create sales objections, cloud the offering to the customer, and obscure the primary benefits. This is a lazy solution and avoids trying to understand the customer's people, who should be constantly seeking new opportunities.

THE STATEMENT OF VALUE

Put yourself in the shoes of the buyer. They want a statement in their—the customer's—language, and the opening statement needs to link you to a previous experience:

1. "Last time you filled your own gas tank it cost you $60; next time, it will cost you only $30 to drive the same mileage."

2. "When you search competitors' websites for information, you can make two, or even three, attempts with a keyword and still not find the information you want. This software will ensure that a search on *your* website takes the customer to your solution the first time they search."

3. "Last time you cleaned the kitchen floor it probably took you 40 minutes with a lot of messy mopping and buckets. This mop will do it in five minutes, and you need no other equipment."

Statements like these grab your attention. You are interested and you want to know more. The value proposition needs to link to the existing solution, with a "similarity statement" so that people feel secure. They will be cautious of having to change their behavior. Remember the previous chapter and the resistance to change.

First, you need a statement like "You can recharge at home and also go to regular gas stations" (this addresses concerns about "refueling" an electric car), "You will use your existing software and it will not be impacted" (this addresses concerns about new software damaging existing software), or "You can store the mop easily" (this gets the user thinking in the context of existing cleaning equipment). The electric car example also points again to the importance of infrastructure for a new innovation.

Then you need the two or three primary benefits the buyer will get from your product. A busy person's memory span is two or at the most three items, and, remember, in selling business-to-business, they are probably going to have to agree on the benefit with someone else before they buy your product. Make the sharing easy for them. List the two or three benefits in each of the two or three categories of time, money, and people. Then develop your master list.

DEVELOPING THE VALUE PROPOSITION

Let's change from the everyday examples I have used so far and imagine a radical new training package that is being offered to your company. The features of this package are

1. A workbook for note taking with "in company" examples
2. A supporting website with real-life examples
3. Training of trainers in-house to deliver training
4. Opportunity to break training from one day to four hours and two-hour modules
5. A workbook in color with photographs
6. Discount for payment for training ahead of time
7. Evaluation of training effectiveness
8. Development of training delivery plan
9. Certification for all trainees
10. Follow-up after three months to address any competency gaps

A general value proposition will probably capture the features that appear to provide benefits in the eyes of the training provider. Clearly, a specific value proposition has to be developed for a particular client.

Starting with the general proposition and where the benefits are most likely, we might find that the attractive features from the supplier's perspective are the website, the color workbook, and the certified trainers because these are the most visible to the suppliers of the training. However, for the company buying the training, the "in company" examples, the provision of training modules, and the three-month follow-up are probably what provide the best value.

Testing features to establish key benefits is something that should start during the development phase, and as a result, the value proposition—with its "pain statement," the similarity statement, and the two or three key benefits—comes together easily at this final stage.

In developing the value proposition we also need to focus it on the risk points in the delivery chain. Remember Michelin's run-flat tire, first described in Chapter 8. They had two main obstacles or high-risk points: repair shops and auto manufacturers. The value proposition needed to focus here as well as on the consumer. The consumer ultimately pays for the innovation, but repair shops and auto manufacturers must be persuaded to install the technology. Look at the high-risk points in your delivery chain and develop your value proposition to fit.

NEW PRODUCT INTRODUCTION

In all the excitement of the marketplace, it is easy to overlook operations issues. We are asking a group of people who produce the product or the service that ultimately creates value for the customer and revenue for ourselves to do something they have not done before. Did someone say *resistance to change*? Of course! If I am being asked to manufacture something entirely new and my production rate suffers and my income suffers, of course I resist.

Do you have that term NPI, *new product introduction*? Do you have a procedure for the pilot test? The product may be different every time, but the procedure is the same, and yet you keep reinventing the wheel. Figure 10.1 is an example of what such a procedure might look like.

Title page

Procedure title: Pilot test

Purpose:
To ensure that a new model or modified design can be manufactured to meet product specifications under controlled conditions.

Scope:
All new models or modified designs

Responsibility: Project engineer
Operations manager
Assembly supervisor
Quality inspector
Auditor

Definitions: SEI; Supplementing engineering instructions
Routing sheet; Description of equipment features

References: Routing sheet
Supplementary engineering instructions 2G-G04-09
Demo test report
Demo test operating procedure 2G-G09-03
Development schedule
Parts list

Records: Pilot and demo test report (retention one year)

Figure 10.1 Pilot test procedure.

1. The project engineer delivers the approved demo test report to the operations manager.
2. The operations manager ensures that a standard cost has been established and a quality plan has been implemented.
3. The operations manager finalizes the routing sheet and forwards it with the demo test report to the assembly supervisor.
4. Any difficulties in meeting lead time are advised by the operatons manager to the project engineer to enable revision of the development schedule.
5. The necessary parts for the pilot test are ordered by the project engineer on the finalized parts list from stores. A minimum of three models are manufactured.
6. The storekeeper delivers the required parts to the assembly supervisor.
7. New parts are approved by the project engineer. The supervisor approves the other parts.
8. The industrial engineer revises work methods and reassigns employees as necessary and advises the time allowed for operatons to the assembly supervisor.
9. The industrial engineering technician ensures that equipment, tools, and templates are installed and operational.
10. The order for manufacture is issued by the operations manager to the assembly supervisor on the development schedule as to date and quantity.
11. Manufacture is initiated and supervised by the assembly supervisor.
12. The pilot test is executed by following regular production procedures.
13. The quality inspector ensures that equipment is clearly identified during manufacture.
14. The quality inspector records any problems experienced during the test.
15. The equipment must be entirely manufactured on the production line. When manufacture is complete, the equipment is audited by the project engineer, the quality inspector, the auditor, and any additional resource person required.
16. The auditor and a draftsman check the UPC number of the completed equipment in the GEAPR system and record these numbers in the pilot and demo test report.
17. Any problems experienced and responsibilities for action are described on a signed pilot demo test report, which is issued by the project engineer and filed by the quality inspector in the test file.
18. Copies of the report are issued by the project engineer to resource persons.
19. Noncompliances are corrected by the assembly supervisor before the equipment is shipped.
20. The project engineer ensures that corrective actions are complete.
21. Pilot equipment is released for sale on the approval of project engineering, the quality inspector, and the auditor.

Figure 10.1 *Continued.*

	Doing it is getting the product to market	Strongly agree	Agree	Disagree	Strongly disagree
1	We get new products to market quickly				
2	We get the ROI we want on new products				
3	Few competitors are able to copy our products				
4	We penetrate all market channels and regions with new products				
5	We withdraw products that fail				
		×4	×3	×2	×1

Figure 10.2 Innovation process assessment.

The pilot run of the new product should be initiated during the development stage to ensure that when you "go live," there is a minimum of disruption in the business.

Now score your organization on how well it performs on this final stage of the innovation process (see Figure 10.2).

A MIND MAP OF THE INNOVATION PROCESS

In order to give you a total picture of the innovation process, I have developed a "mind map" of the process, shown in Figure 10.3. The map shows each of the significant activities at each stage in the process. It does not capture everything, but is a useful checklist.

LOOK FOR OTHER MARKET OPPORTUNITIES

IBM's Global Expense Reporting Solutions were developed to automate the company's internal travel booking and expense reporting. IBM found that

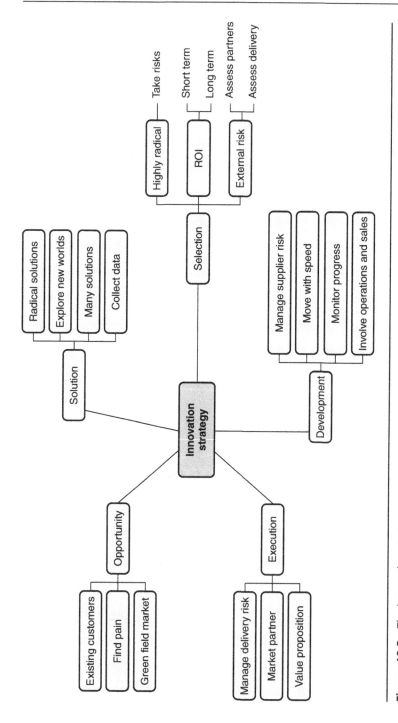

Figure 10.3 The innovation strategy.

it reduced administrative costs by 60% to 75%. Realizing that many of its customers would be interested in achieving such savings, IBM has provided the service to organizations worldwide, creating a new business.[2]

YOUR INNOVATION PROCESS CAPABILITY

The maximum score in each stage of the innovation cycle is 20, and adding your five scores gives you an overall percentage score. If you scored less than 50, then your organization is not in the game. If you scored over 60, but one of your areas is below 10 points, then that area needs serious attention. Look at the activities I have described in that area and identify where you need to act.

Now you have an understanding of innovation strategy and process. Let's look at the cultural aspects of innovation in more detail.

KEY POINTS

- The best ideas don't always make it.
- A fast and reliable delivery system is essential.
- Engaging operations and sales people in the development stage prepares them for the delivery stage.
- Competitors with inferior offerings will copy you if you are not fast to market.
- The customer value proposition should be confined to two or three key benefits.
- It is tempting to describe the features that have been the focus of the development stage.
- The customer's primary interest is benefits, not features.
- The benefit statement should link to the customer's previous experience and then show two or three advantages of your new product.
- It may be necessary to develop a different value proposition for each buying point in the delivery chain.
- The organization must have a highly specific NPI (new product introduction) procedure.

Part III

The Culture

11
Culture and Behavior

He who innovates will have for his enemies all those who are well off under the existing order of things, and only lukewarm supporters in those who might be better off under the new.

—Niccolò Machiavelli

Machiavelli was the first person in modern recorded history to write about innovation. His chilling words are a reminder that we have to plan very carefully the change to an innovative culture. There will be many who resist either overtly or covertly. I will discuss change in Chapter 19, but in Chapters 11 through 15 I will look at the various aspects of culture, and in this chapter focus on the behavioral component.

We all struggle with what exactly we mean by "company culture." It is typically defined as the attitudes, experiences, beliefs, habits, and values shared by the people inside and the stakeholders outside the organization.

Behaviors are based on values, in other words, what people "value," or see as important. Deal and Kennedy described corporate culture as "the way things get done around here."[1] The world of ISO gives me the privilege of seeing many cultures around the world. The subtle differences between the cultures of different nations never fail to intrigue me. Cultures are very much the result of a nation's history. Company culture is the result of a company's history.

The three cultures I know best are America, Britain, and Canada. I have learned a lot about the Latin cultures of France, Spain, and Mexico, and I am fascinated by the Eastern cultures of China, India, and Japan, especially as reflected in their art.

America's century of success as an innovator has come as a result of its diversity, driven by the famous words from the Statue of Liberty, "Give me your tired, your poor, Your huddled masses yearning to breathe free, The wretched refuse of your teeming shore."

Britain, on the other hand, does not handle diversity well. A thousand years of repelling invasion means the British treat foreigners with skepticism. On the other hand, Britain has always networked well, whether in its pubs or in its educated elite. Again, it has a strong history of innovation.

Canada is interesting. It has the diversity of the United States, which gives it a culture of tremendous creativity, with a population only the size of California. However, unlike the States, it often fails in execution. It does not switch well from loose to tight.

There are many types of culture; Deal and Kennedy describe, among others:

The macho culture, with quick feedback and high rewards. You think of the stock market and sports teams.

The work hard/play hard culture, which has replaced the macho culture in terms of "fashion." This culture is the customer service culture, with its own private language and focus on team. It is seriously risk averse. This is the way of the "organization of the '90s" and does not lend itself to innovation.

The process culture, in which people are task oriented. They produce consistent results, but always the same, and innovation is again a nonstarter.[2]

You hear the terms *strong culture* and *weak culture*, and there is a tendency to think *strong* means *good*. Strong actually means *cohesive*, but if this produces alignment with the wrong values, it will not produce innovative outcomes for your organization. The strength may not produce innovation. *Values* are the beliefs and ideas that people see are important. A weak culture is of course one where the values of people vary, and things only get done through procedures.

Both strong and weak cultures have downsides. The strong culture may suffer from groupthink, where a comfortable consensus overrides the tough decisions. Nobody is prepared to challenge the status quo. The lack of challenge outside of an immediate task means innovation is a nonstarter. On the other hand, the bureaucracy of a weak culture kills innovative thinking at the outset.

What you will see is that the culture of your own organization is a mix of behaviors and beliefs that have made things work for you in the past.

We all know the old saying, "If you do what you always did, you will get what you always got." If we are not an innovative organization at present, then clearly something has to change. Unfortunately, cultures are very subtle, and change is not easy. We have to ask, "What are the beliefs and values that must change?"

The sadness of much of quality management is that it has lost the focus on the customer and too frequently has focused on internal efficiency, using the customer as the excuse for driving that efficiency. This is short-term thinking and, in truth, focuses on the short-term interests of the shareholder or business owner and not the long-term interests of the organization.

Long-term thinking is interested in what today's customer needs tomorrow, and who will be a new customer tomorrow.

Quality management has also focused too much on process and procedure, and explicit, or documented, knowledge. The beliefs and values in an innovative organization will need to shift to a better balance of people and process, and a greater emphasis on the tacit knowledge I described in Chapter 2.

There are a number of values, and hence behaviors, that an innovative organization must develop. They all stem from the need to develop a learning organization.

BEHAVIORS ARE BASED ON VALUES

The behaviors that most organizations need to develop in order to become more innovative are:

1. Exploration
2. Interaction
3. Observation and note taking
4. Collaboration
5. Experimentation
6. Ability to embrace failure

EXPLORATION

Exploration is a skill for both the creator and the connector. This is how we find opportunities, but also, as with Henry Ford, this is how we find solutions.

In today's business world, exploration is the victim of "good time management." We have become obsessive about process efficiency in our operations and have extended this mind-set to all aspects of our business life. Archimedes's moment of "Eureka!" occurred when he was having a bath. Your best ideas will come when you allow them instead of forcing them. We must allow time to "bathe" and relax.

Google used to be famous for their "20% time." They let their staff "explore" one day a week in areas that were not directly project related. That led to the creation of Gmail, AdSense, and other Google innovations. However, in August 2013 word got out that they had stopped doing this. Senior management had created productivity goals that effectively eliminated the 20% time. The questions on the table became "Would Google lose their famous creativity over time?" "Had short-term results overtaken long-term vision?"

Allow yourself, and your teams, to step out of the box and explore. Leadership teams have retreats. If we are going to release the collective subconscious, *everyone* must step out of the box.

Some of my best ideas ever have come from chatting at the bar in rugby clubs. I do my best writing in coffee shops while I am people watching. I choose to live out of the city so that I can release the learning that has accumulated during the interactions of the day in the city.

My innovation workshops are at a beautiful location called Kingbridge. Everyone loves their day out of the box. They feel the ideas coming free.

Go to places you haven't visited before, meet people you haven't met before, experience things you find a bit intimidating. Remember the old saying "travel broadens the mind." You should join a network or club. There are many around these days. I don't mean web-based, I mean face to face. Gen Y have become notorious for not being "joiners." If you are Gen Y, give this some serious thought.

I am fortunate to be deeply involved in ISO, and that gets me traveling many places. I always take an extra couple of days after my meeting to experience the culture, whether it is the street market in Nairobi, a police chase in Kuala Lumpur, a mummy's tomb in Egypt, or Dracula's castle in Transylvania. You don't have to travel the world, by the way. The world is in your own back yard! I drove across Iowa and developed a large chunk of this chapter at the Texas Grill in Des Moines. New experiences broaden your perspective. They also enable you to see new opportunities. You get these opportunities to travel. Take them. Then allow yourself time to step back from your experiences. I mentioned in Chapter 1 that I have that fortune cookie note that should be the mantra for every innovator: "The secret of a good opportunity is recognizing it."

INTERACTION

Mix with interesting people. This may surprise you, but I do not mix easily. Put me in front of a thousand people to share my life with them and I have no problem. Put me in a room of 50 garrulous and assertive people at a cocktail party and I go quiet.

If you are like me, let me share a tip. If you want to find someone interesting to talk to at a cocktail party, a great technique is to relax by a table, visualize your favorite room at home and that the people in the room are all your best friends. In five minutes this visualization will draw to you, almost like a magnet, someone in the room who thinks you look like an interesting person.

A great book on networking technique that I can recommend to you is *Never Eat Alone* by Keith Ferrazzi.[3]

OBSERVATION AND NOTE TAKING

This skill goes hand in hand with the exploring and interacting that creators and connectors must do. Capture the random thoughts that flash through your mind as you explore and interact. Don't "lose that thought" into your subconscious.

Always have a small notebook with you. Have a book that fits in your pocket and is a book that you are proud of. I bought mine at a store called Indigo. It has a leather cover with beautiful embossing. The pages are perforated and will tear out without damaging the book. Writing in the book is a wonderful experience for me.

However, life doesn't always work like that. You may get caught without your book. I was sitting in Charlie Biddle's jazz bar in Montreal listening to Charlie himself on the bass. I had been struggling for months to find an opening chapter for my first book. My thoughts suddenly came together in my mind, and I had no paper. The lady at the bar gave me a white paper placemat and a cheap pen. My opening pages of *Do It Right the Second Time* were written on a paper placemat.

Charlie finished his set and, walking past me, thought I was writing a song. I won't forget the words of the late Charlie Biddle. He said, "Write one for me, man."

Make a point of traveling in different cultures to get new experiences. Innovators always take notes. You should always be learning.

Observe with an open mind and ask why things happen. Don't judge. Find out about other people's lives. Too much of business training is analytical and judgmental. Write notes later, but not too late. The ancient practice of keeping a diary is exactly this. Always be looking for what bugs people, but do not try to solve their problem; simply find out what their problem is. Solutions come later. And if you want the real truth, talk to kids.

One final piece of advice—always travel on Robert Frost's road less traveled.

COLLABORATION

This skill is key for the connector and the developer. It becomes essential as we work on conceptual solutions and also as we develop working solutions. "Two heads are better than one," and collective knowledge produces powerful solutions.

To develop this skill, find three or four other people that share a passion for your topic. This is networking, but more focused. This is the beginning of a *community of innovation*, which I describe in Chapter 12. I collaborate with people in Europe as well as in North America. Tools like Skype make this so much easier these days.

ASQ formed its Innovation Interest Group on the journey to becoming a division. We told people and they flocked to join. I gained enormous learning from talking with people at the Group conference and exchanging random e-mails. I made new friends, and that matters in the world of innovation.

Freely exchange your knowledge. When others send you articles or ideas, take time and space to read them and see what ideas they trigger for you. Every six to eight weeks I do a Skype call with my friend Chris Hakes in Europe. The interaction always generates fresh perspectives. In between, we fire off articles to each other that we think will be interesting. I talk more about collaboration in Chapter 14 on networking and Chapter 18 on open market innovation.

EXPERIMENTATION

This begins with the work of the connector, who must test alternative concepts, but becomes essential for the developer as we try out working solutions. Experimenting means being prepared to fail. It also means being surprised with an unexpected outcome. Post-It notes came from a failed experiment, and the name WD-40 refers to the 39 previous failures.

WD40 is short for "Water Displacement, 40th Formula." After many failures, Norm Larsen "invented" WD-40 in 1953 in San Diego, and over 60 years later it is still regarded as one of the most useful products on the planet. What drove Norm Larsen to create WD-40 was a "need." WD-40 was not a solution looking for a problem, it was the answer to a need at the time. Auto ignition systems failed more often due to shorting of damp plugs and cables than any other single cause. Norm's question was "How can we drive away damp." He answered that need, and his product went commercial in 1958. After that, nobody left home without WD-40.

When you prototype, use techniques like storyboarding that are tactile and interactive. Use tools like, yes, Post-It notes, that again are tactile and interactive.

When I teach process mapping I always use paper and Post-Its. I refuse to use the computer screen. It is too controlling and excludes 80 percent of the participants. On the other hand, use the technology of video extensively. This is a wonderful learning tool. Your culture must accept unpolished presentations. Polished PowerPoint presentations exclude too much of the original thinking.

One other key point: involve your customers, managers, and business partners continuously in the experimenting, and always have multiple prototypes to show. An important part of experimenting is accepting failure.

ABILITY TO EMBRACE FAILURE

More than anything else, innovation is a "risky business." Innovators are not afraid to fail. *Risk* is the four-letter word that stops people from innovating. In large part, we are averse to risk because we understand it far more than we did a decade ago. We are able to calculate risk, prioritize risk, and mitigate risk. Because of all these abilities, we have developed a mind-set that risk should never enter our lives.

The "taking care of number one" mentality, which frequently devolves into a corporate culture of mistrust and backstabbing, leads to people being inherently skillful at distancing themselves from failure. People see a project failing and they stop attending meetings and stop returning phone calls. A project finally fails, and there is no project closure or lessons learned session.

It's like not doing an analysis on your customer complaints. These are one of the richest sources of knowledge.

One of the most common failings I see in project-driven businesses is the absence of project closeout discussion, analysis, and documentation.

The excuses are always "we ran out of time" and "people got moved to other projects." You must learn from experience.

I am now going to add a seventh behavior to the list.

RECOGNITION OF BEHAVIOR

You must recognize new behaviors, not results. Recognize innovations and not heroic defenses of the old. Endorse the behaviors I have described. Reinforcing new behaviors builds trust, especially when they are based on the new values that they imply. One other point—you must move eventually to a point where recognition and reward are aligned. Phil Crosby, in the 12th of his 14 steps, said, "Appreciate those who participate." I talk more about recognition in *Do It Right the Second Time*.[4]

CHANGING BEHAVIOR

So how do we begin this change of behavior and culture?

The challenge is clearly one of *Leading Change*, which is the title of John Kotter's classic book on the subject.[5]

There are a number of issues that will create resistance to the change to an innovative culture. You see these in the new behaviors I have described and also from looking at your existing culture. The primary change agents will be your change team, and I will describe the change process more fully in Chapter 19, "Lighting the Fire." Resistance will be especially strong if you have a history of success. It will be necessary to create a sense of the need to change. The process for handling change should follow this approach:

- *Create a sense of urgency.* Create a "burning platform."
- *Create the change team.* Their mission is to quickly create a critical mass of believers.
- *Early win.* Produce a quick result, which is then publicized.
- *Communicate vision.* This becomes a key task for the business leader.
- *Enable action.* People must be given the authority and time to innovate.

- *Don't declare victory.* The short-term win creates danger as people relax.

- *Anchor new approach.* Embed new behavior until innovation becomes business as usual.

- *Recognize success.* The behaviors described earlier are endorsed.

This gives you an outline of the change process, which will be detailed in Chapter 19. The business leader plays a key role in all of this.

MAINTAINING AND STRENGTHENING VALUES

Whether you call it *performance management* or *annual assessment*, the one-on-one performance review is one of the key activities that will either endorse or undermine the culture you want. If you want to be innovative, the goals that people first agree to and are then assessed on need to reflect the values and behaviors of innovation. The people conducting assessments need to fully understand the meaning of innovation. This is where you take time to see how much exploring and note taking got done. Who were the collaborators, which experiments succeeded, and which ones failed? This is where you congratulate people for having the courage to fail. If these behaviors are listed in the review input and again in the half-year assessment, they will start to be taken seriously. This will also develop the other key components of an innovative culture: trust and honesty.

RECRUITMENT

The other process in which you can unknowingly undermine your culture is your recruitment process. An innovative organization embraces diversity. There is a tendency for us to recruit more of the same when we look for new people to join our organization, and we mirror the majority of our company population. Your recruitment focus should be on the values that matter in your culture. You must evaluate whether potential new people support those values and practice the behaviors you seek.

When I joined the Research Division of the Courtaulds Corporation, the company had each potential employee spend three days in interviews. This was the way in which they ensured that new people had the right values. Compare this investment with the cost of labor turnover. Companies are often surprised that when they invest in only a one-hour interview, they also have high labor turnover.

> **KEY POINTS**
> - Culture is "the way things get done around here."
> - Strong culture means cohesive; weak cultures have variation.
> - Culture is based on values, or what is important.
> - Cultures such as "macho," "work hard, play hard," and the "process" culture do not support innovation.
> - An innovative culture has to be strong to be effective but must be based on unique values.
> - Behavior is based on values; innovative behaviors are exploration, observation, interaction, collaboration, and experimentation.
> - Embracing failure is a critical behavior.
> - Recognition endorses values and behaviors.
> - Values are maintained and strengthened through the performance review process and through a sound recruitment process.
> - In order to change to an innovative culture, an organization will need to go through a distinct change process.
> - The business leader has a key role in communicating values through a vision of innovation.

12

Innovation Teamwork

No man is wise enough by himself.

—Plautus

A rugby team is really two teams: the *forwards* or "scrum," which are eight "big guys," and the *backs* or "three quarters," who are the six "fast guys." These two teams are joined by a cheeky little guy called the "scrum half." The scrum half is the critical link between the forwards and the backs and spends much of the game constantly chattering to both backs and forwards.

In *Do It Right the Second Time* I talked about how as captain I gave the leading of the forwards to a bearded mountain named Andy Alltree and the leading of the backs to an Irish wizard named Brendan Webb. I played on the wing because speed was my asset, and it gave me the chance to read the whole game, view the opposition's strategy, and guide the team if we were losing our direction.

You can see how these roles play out in your own innovation teams and how this manifests what I call "de Bono's law," that the natural span of communication is six to eight people. A lot has been written about the experience of running teams of 20 to a hundred people, but I will come to that later.

I must confess that, for myself, it has not always been easy to move beyond the old teamwork methods of business. My experience in both sports and in volunteer work such as ASQ has been a great help. The ASQ Innovation Division is a very large team, and we had a core team that set goals. We share the leadership, and one of our successes has been our low-budget conference, which is in itself an innovation. First year in Sacramento, second year in Toronto, and third year in Charlottesville, Virginia. We sacrificed the sacred, created a really good conference with a registration fee of only $199, and enabled all in the group who had the desire and ability to speak at the conference. The model was not created by one person overnight; it was the outcome of collective knowledge.

There are essentially three types of "teams" you will experience in the world of innovation:

1. *The community of innovation.* This is not like a normal team; it is much looser and is focused on sharing and growing knowledge. I will explain how such a community operates.

2. *The project team.* This is probably something you are quite used to, although innovation project teams do have some unique attributes, which I will explain in this chapter.

3. *The change team.* This operates at an organizational level. I briefly explained its role in the previous chapter when I described changing behavior, and I will explain this in more detail in Chapter 19.

COMMUNITIES OF INNOVATION

The famous solution to this problem of bringing people together to be creative is often called "skunk works." Lockheed in the 1940s created the term when they named their community of innovation after the moonshine distillery in the cartoon strip *Li'l Abner.* Skunk works is now a widely used term. Skunk works often operate without the knowledge of senior management.

The term *community of innovation* is the product of knowledge management and is a fancy name for skunk works. They are multifunctional networks of relationships, and don't look like teams. In networks, information becomes knowledge, and the best innovation comes from organizational knowledge. It is the communities of innovation that enable the best knowledge transfer. I will discuss networking in a lot more detail in Chapter 14.

Trust is essential for the community to work, and they recognize that learning is a social activity. The volunteers have a passion for the subject. Communities are "coordinated" rather than "led."

The coordinator of this community must have a passion for people, provide resources and tools, and also introduce new members. The coordinator monitors the big picture, such as number of e-mails and level of participation, and identifies the issues to be addressed.

Community members have a passion for the subject, and it is a game of passion and friendship, and all newcomers are welcome.

In forming a group, we must be clear on its purpose, and we must also be clear on the type of contribution each person can make. We may give a person a role, but they may demonstrate strength in another area over time. Your initial role in the community will be as a creator, connector, developer, or doer.

As you become more confident in your ability to develop collective knowledge, include your customers and suppliers and move into open networking. There is even more knowledge to be gained. Many feel nervous about open networking, but you can be certain someone somewhere has seen the same opportunity as you have. We had a saying when I worked in the textile industry that there are no secrets. The ones who do it best will win.

ASQ operates a number of interest groups that are knowledge-sharing communities. The Innovation Interest Group (IIG) was formed in 2013 and morphed into a Technical Committee on the way to becoming a division. The IIG had a variety of "meeting modes." Initially, there were just planning activities, but over time there were face-to-face meetings in a conference environment, and webinar sessions providing members with knowledge on specific topics. In between that, members shared information by e-mail. Periodically, the "core team" would meet by phone to plan forward events.

These communities are very much about sharing and growing knowledge and, for a company looking to initiate innovation, provide a "seeding ground'" for an innovation project. A community of innovation will likely produce an informal opportunity, or even a potential solution to an existing problem. This leads to a team focused on a specific opportunity and the creation of an innovation project.

INNOVATION PROJECTS

When you create your project team, you need to know in advance who are the creators, connectors, developers, and doers. Try to build a good mix, and of course an overriding attribute is a passion for your subject. The blend of these four attributes will shift as you move through the innovation project. At the first stage of the process, there will be a majority of creators. At the last stage, a majority of doers.

As a project team, you will initially operate in a loose mode, and there must be a predominance of creators and connectors who will see opportunities and find solutions. They will network with others both inside and outside the organization. Remember Linus Pauling: "The best way to have a good idea is to have a lot of ideas." The techniques of ideation I described in Chapter 6 are what you use to develop these opportunities into solutions.

LEADING A PROJECT TEAM

An innovation project needs to operate at both a strategic and a tactical level. The strategists set initial direction and reengage at the tipping point of the project. The tacticians operating the project need to be given a free hand, especially in the early "creative" phase of the project.

Linkages need to exist between the team and the organization in which it operates. An organization should allow a "network" structure in the team to enable an innovation project to succeed. The organization creates context for the project; the project creates context for its people. The parent organization must enable the flow of knowledge and allow the project network to function. Project leadership should report to a higher level than is normal in a parent and hold separate meetings from the parent. The parent must allow the network to exist, increase autonomy, and encourage diversity.

Having identified who the right people are, the next issues might sound surprising. We need to really know who's on the team. This is often unclear at the start of a project. Be ruthless about membership, and provide clear challenge based on the project charter, even at the initial creative phase.

If you are leading the team, the first meeting, even the first few minutes, are critical. Experience shows that we need to start the work by being task oriented and assigning roles. Don't worry about the approach; it's more important to get an early win. Relationships inside the team will build by doing the work. Set "reviews" and "lessons learned" sessions and be constantly mentoring. In the early creative stages, trust is difficult because people are new, but trust is essential. The leader "holds the team together."

First, face-to-face meetings are imperative. It's not just about transferring information; it's also about building relationships. Trust is based on behavior and being able to rely on others. This only truly develops in a face-to-face situation. Don't slip into the strictly conference call mentality.

Secondly, new people must be mentored into the group, not through a formal HR procedure but through informal and personal attention from the leader. A one-on-one meeting or a personal phone conversation, if geography is a problem, should be held prior to the main meeting.

During the main meeting it is essential to talk off-line and ensure that the new people are truly engaged. The gift of time is one of the most precious gifts from any leader and is truly appreciated.

Thirdly, and this will be surprising, research has shown that in the early stages of the team, roles need to be well defined, while at the same time, the approach should be loose. Many think that leaving roles vague encourages sharing, whereas in practice, having a defined role makes people feel relevant and causes people to feel that they can work independently and contribute to the group outside of meetings. Another research finding is that in the creative mode, having a loose process leads to people investing more time and energy in collaboration.

You can see that a number of these actions seem to work against each other, and so the job of a leader or facilitator is complex and requires balance.

When the group reaches its tipping point, it also reaches the toughest job of all. The mode of operation changes from loose to tight.

To provide an example of sharing roles, the International Working Group (WG) that developed the standard ISO 10018, *People involvement in management systems*, and which I chaired, comprised delegates who have experienced firsthand the problems from not involving people in the running of a business.

There were some critical tasks to be performed in our meetings as we created a radically new standard. We had a good sense before meeting of where people's aptitudes lay. My own task was to make sure we got to "end of job." Our finished document had over 6000 words. Research has shown that too much shared information becomes a problem for a group over time. The role titled "secretary" was really the task of information manager. We broke each week into 90-minute segments, and Paul Simpson from the UK acted as timekeeper and was time manager. Renato Lee from Brazil was language monitor. Renato ensured that everyone understood the terms that were used in conversation. The delegate from Sweden challenged the team if we were galloping ahead and missing something important. All of these roles were shared, and there was overlap. We were a diverse group with different native languages, but our working language was English. A good team does not compartmentalize the roles of its members. At the same time, people take responsibility for their own roles.

EARLY RESULTS

In these early stages the leader has an interesting challenge. You would think that the challenge is to develop the network and the relationships so that people work well together. In truth, there is the bigger challenge of showing results. An early win is vital. The leader has to be process-focused and ensure that there are documented outputs even at the earliest stages. If not, the team will lose its passion and rapidly disperse.

Still in loose mode, the team will start to capture some potential solutions. In the early days of the people group that I mentioned, the task was the creation of a first working draft (WD1). This WD1 listed problems they had encountered through lack of people involvement in management systems. Inevitably, we would find potential solutions but we would not discard those solutions. We would record them.

There will be a point, though, when a project team morphs into *problem solving* or *connecting* mode. This is where the team numbers are likely to balloon, and where, from a team perspective, new challenges arise.

In the *problem solving*, you need diversity and disruption. That's how you get the wild ideas. The person on my standards writing team who had a different objective provided the diversity we needed. This way, we challenged ideas at the later stages of this creative phase. We were writing a completely new standard with no precedents, and challenging ideas forced us to explore widely and deeply.

In the early stages of creativity, you should allow social time. The more analytical people may think this is slacking, but this early socializing is when you build trust, even though you also have diversity and disruption. In these early stages, your goals should be short term, giving you those essential early wins. This may again make analytical team members feel uneasy. There will be those who want you to tell them what the final answer will look like, but you must tell them bluntly that you have no idea because if you did know, you would not be innovating.

COLLABORATION

Work by Gratton and Erikson confirms our own empirical experiences.[1] As teams grow beyond the 20-person mark, and also as teams become more virtual, collaboration decreases unless it is structured into the work. Complex teams are also less likely to help one another.

The factors that inhibit collaboration are nationality, age, education, and especially length of time on the team. In the world of ISO standards, I see the first three issues addressed very positively, and they have become a great asset in creating standards that are accepted internationally. However, I often see a group that started a particular piece of work feeling threatened as new people join the group. It is important to remind ourselves that more people means more ideas and more solutions.

One other thing to beware of: the greater the proportion of experts on a team, the greater the likelihood for conflict and stalemate.

Overcome these problems by allowing space for people to relax and chat. This builds trust as people learn how the other person operates. Find goals in common and work on those at the start. Most importantly, find early wins.

However, large group collaboration can work with the right commitment from its people. Linus Torvalds completed the final and most difficult task in developing Unix software, and as a result it became known as Linux.[2] Many, many people in a collaborative web did the preceding groundwork. Silicon Valley succeeded where the Boston-based Digital Equipment, Wang, and Data General died. The Silicon Valley folks hung out in bars and coffee shops and freely shared their knowledge. It's no surprise that Starbucks was conceived on the West Coast. In Boston, people got sued for sharing.

PROJECT CHANGE: LOOSE TO TIGHT

This change is not sudden. I talked about this in Chapter 8. A project team should prepare for it. They will have been looking at their list of potential solutions and assessing risk and usability. The list is condensed down to, say, six or at the most 10 preferred options. The tipping point is where the group reconnects with business strategy and two or at the most three "hot options" are selected.

Moving forward from here, the team changes. Some of the highly skilled creators and connectors will leave. They go back to looking at new opportunities. New faces will have joined at the problem solving stage. Depending on the numbers you have, you may need more developers to join the group at this point.

If you are a small or start-up business, your challenge is much greater. Your team is going to comprise the same people, and so the behavior change in the team becomes much harder. Conversely, the project management for a small business is much easier.

TEAM CULTURE SHIFT

The challenge for innovation is that we need two cultures. Many of the behaviors are the same in the two cultures, but there is a fundamental difference between the creative phase 1, where there is a need for freedom of thought, and execution phase 2, where there is a bias for action.

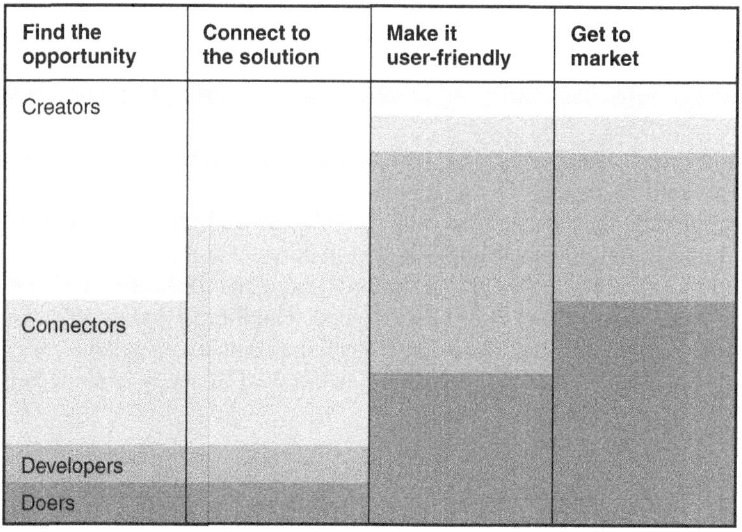

	Find the opportunity	Connect to the solution	Make it user-friendly	Get to market
Creators				
Connectors				
Developers				
Doers				

Figure 12.1 The shift in team content and culture.

Many organizations have solved this problem by separating the two cultures into two distinct parts of the organization. This seriously impedes the communication down the innovation chain. Far better to transition the culture between each stage of the process (see Figure 12.1). You can do this by knowing the kind of people you have and grouping them at each stage in the process so as to encourage the behavior you seek

At the development stage, speed becomes important, so you will probably break the group into a series of task groups, each addressing distinct aspects of the solution. The solution is now starting to become a new product or new service, and each of these task groups will address a component or element of the product.

HANGING TIGHT

At the development stage, the challenge is to develop usability of the new product, and the group will engage customers to assess whether the behavior change required by the product or new process is too extreme.

We now reach the point where we reengage with the senior leaders and strategists of the organization. Avoid providing them with the raw data for decision making. Instead, analyze data to get information and knowledge to help your senior leaders make informed decisions.

You will probably be reengaging more closely with the project sponsor during this time. I encourage you to build that into a strong relationship because you will need their close support in the execution phase. That will be when the politicians and people with vested interests emerge and attempt to derail the project.

Your picture of the group should now be shifting. Instead of an amorphous, free-flowing mass, it has now segmented into a set of tight fighting units. Each unit probably has only five to seven people.

Each of these fighting units will be highly task oriented. They will have very specific goals. The goals will be SMART, or should I say "SCHMART," that is, *specific, challenging, measurable, agreed, realistic*, and *time bound*. This may not be new to you. One task may be sourcing critical suppliers or subcontractors and setting up contracts. Another will be refining a component or subprocess. A third will be testing alternatives on the early adopter. The leader has to keep these task groups aware of each other's work. Frequent huddles are essential, maybe even a couple of times a week or more.

The new product and its components are being tested, and the overall group is working to a tight project plan. This is a familiar mode of operation for anyone familiar with an R&D environment. The task groups at this stage should be keeping concise and understandable records. Communication between task groups is hard. Records should not be long dissertations. They should be crisp and easy-to-read notes. Networking between teams is also critical, and in a large group each task group should have a liaison person whose job is to scan the work of the other task groups.

DELIVERY

At the final delivery stage, the team changes again. You will keep a creator and a connector and certainly a number of developers, but here is where operations and sales people become the major part of the team. These people are very task oriented, and so the team culture will shift once more. The sales people gain feedback from the marketplace, and the operations people from manufacturing and service delivery. This enables the developers to further refine the product.

You have seen from this chapter and the previous chapter that a number of key competencies are required. In the next chapter we will look at the attributes of the competent innovator.

KEY POINTS

- Innovation succeeds through the sharing of group knowledge.
- The natural span of communication is six to eight people.
- *Skunk works* is a "nickname" for the *community of innovation*.
- The coordinator of the community must have a passion for people, identify key issues, and provide resources.
- As a community grows in size, collaboration becomes more difficult.
- The project team will start in loose mode, but it is essential to achieve an early result.
- A high proportion of experts impedes collaboration.
- The leader or coordinator must ensure (1) face-to-face meetings, (2) mentoring of new members, and (3) clear role definition for team members.
- A project team changes from a loose to tight mode of behavior after the tipping point.
- Team membership may shift after the tipping point.
- For a small business where membership can not change, behavior will need to change.
- In a larger organization, at the development stage, the project team may break into task groups.
- The project team must involve customers and suppliers at the development stage.

13
The Competent Innovator

*I keep six honest serving-men
(They taught me all I knew);
Their names are What and Why and When
And How and Where and Who.*

—Rudyard Kipling

Most people acquire their competencies through that terrible thing called *experience* and not through formal training. I had a boss once who said, "The worst thing about experience is getting it." Education and training provide the initial framework for developing competence, but the experiential learning that follows is where the skills and abilities are developed.

The word *competence* is cumbersome, but it is fashionable, so I should use it. I much prefer the five-letter word *skill*, but that word has come to be applied more in a physical environment than in a thinking environment. The terminologists in ISO 9000 do a good job on terms and definitions, and their past convener, my friend Bill Truscott, would be thrilled to hear me say that his team has given us a definition for competence of "demonstrated ability to apply knowledge and skills." This chapter will explain how to develop knowledge and skills, and then how to apply them.

LEADERSHIP WORKSHOP

Everyone will first need an understanding of what innovation actually is. The chances are, your own perception has changed since you started reading this book. Start with your leadership.

A leadership workshop addressing the content of this book enables leaders to participate in discussion and address their concerns. Your leadership must understand what innovation is. There are many myths and legends. They must understand the difference between creativity and innovation, and that innovation is not magic and there is a process. Everyone has a role in the process, and the outcome will be achieved through collective knowledge and not the lone genius. Finally, the innovative outcome will not be some sudden epiphany, but will be the result of following the process.

In educating the leadership, I would encourage you to focus in the morning of the workshop on *culture* and *creativity* and in the afternoon on *strategy* and *execution*. The primary concern of leadership will be setting and executing strategy

MANAGEMENT WORKSHOP

A number of the skill sets described in this book are introduced in the workshop description in Figure 13.1. In the morning it focuses on the individual, and in the afternoon on the process and the organization. People participate in exercises and assessments.

After a brief introduction explaining what innovation is and what it is not, I ask each member of the group to answer two questions. First, "What is your primary concern about your organization's ability to innovate?" and second, "What product or service causes your customers the most frustration or difficulty?" This gets everyone involved in discussion, and they are fairly easy questions to address.

Engaging the individual and showing them the value they can personally provide is essential. The group carries out the self-assessment you saw in Chapter 3 and finds out whether they are creators, connectors, developers, or doers. They also carry out an exercise in creativity that emphasizes the importance of interacting with others. While on their own, they may only find three or four solutions to a problem; through the process of ideation they can find 15 to 20 ideas.

We then start to address strategy and how to assess and mitigate risk.

The afternoon moves to looking at the type of organization structure that enables innovation to flow more easily. The group does an organizational assessment to identify where their individual organizational weaknesses lie.

The day concludes with the management team initiating its own priority plan for deploying innovation into their organization.

Introduction. Innovation is the only strategy by which an organization can differentiate itself in the long term. Innovation is driven by a customer's desire for more convenience. "Conventional thinking" and "fear of failure" inhibit innovation.

Workshop: "Assessment of Innovation Concerns and Opportunities"

Innovative Culture. You learn how to develop the behaviors of an innovative culture: exploration, collaboration, and experimentation. Participants identify how to make their own best personal contribution to innovation, whether by generating ideas, finding solutions, making solutions practical, or implementing solutions.

Workshop: "How You Will Make the Best Contribution to Innovation"

The Creativity Challenge. Participants learn how to work with customers to identify opportunities. Defining the problem and identifying roadblocks and new ways of generating ideas is the path to the conceptual solution. It is shown how a "loose" and open network is necessary for finding radical solutions.

Workshop: "How to Identify Innovation Opportunities and Solutions"

Strategy. Integrating innovation with the strategic planning of the organization is essential. Do people truly want our new idea? How do we select the ideas into which we invest resources? You learn risk assessment and risk mitigation for new ideas.

Workshop: "Risk Assessment and Risk Mitigation"

Executing the Strategy. The focus now narrows. For every 3000 ideas, only one makes it. You learn the importance of managing risk and evaluating the behavior change required in using a new product or service.

Workshop: "Importance of Changing Culture from 'Loose' to 'Tight' in Order to Execute with Speed"

The Innovative Organization. You learn about the different kind of organizational structure that is needed to lead to the creation of breakthrough innovation. You reduce hierarchy while increasing autonomy and diversity.

Workshop: "Evaluate Innovative Strengths and Weaknesses of Your Organization"

The Path Forward. The road map for developing innovation is explained. The importance of the "burning platform" and the "short-term win" are shown. The method for obtaining a short-term win is outlined. These are first steps to creating an innovative organization.

Figure 13.1 Innovation project tools and metrics workshop.

INNOVATION COMPETENCIES

The competencies you need to develop in each stage of innovation are:

1. and 2. *Find the opportunity (creators) and solution (connectors).* Exploration, interaction, observation

3. *Selection of solution.* Risk assessment and mitigation

4. *Development of solution.* Experimenting, embracing failure, project management

5. *Delivery.* Value proposition, storytelling, strategic selling

Networking, note taking, and collaboration apply right through the process.

The primary attribute for the innovator is a "yearning for learning," so we must develop a thinking workforce that is constantly looking for a better way and is willing to get ideas from anywhere. This leads to other competencies, such as an ability to interact with others and willingness to suspend your own judgment.

As I take you through the different competencies, you will see that they overlap. I will discuss these in turn.

EXPLORATION AND INTERACTION

This may not seem like a "competence," but more of a behavior, and yet in today's world we have a tough time doing this. We do have to train ourselves to explore. What you really need here is a plan, both for yourself in your personal life and in your organization. However, be ready to break the plan! Let your emotion lead you and not the plan. I will explain more about this in Chapter 14. General Eisenhower said, ". . . I heard long ago in the Army: plans are worthless, but planning is everything."

Innovations happen at the intersection of disciplines. Dr. Niall Ferguson at Good Hope Hospital has told me how this has become fundamental in the medical profession. A good physician will not just use their own knowledge, but integrate it with the knowledge of the patient and the knowledge of the patient's family. The problem may reside in one area of knowledge, and the solution may reside in another. Innovation takes place in the space between people. The Internet allows people to share ideas, but most breakthroughs come when you are face to face.

This skill is one you use when you interact with customers and suppliers. This is another skill you can develop on your own but can be used both at conferences and in day-to-day activities outside of work.

QUESTIONING AND LISTENING

Remember Peter Falk as Lt. Columbo? He wasn't afraid to ask dumb questions. Every case was a learning experience. Columbo was an auditor at a crime scene. Good auditors are not predators; they are people who want to learn. Remember, the innovative organization is a learning organization. The innovative person is a learning person.

When I train auditors, I always share the quote from Rudyard Kipling in his book *Kim*:

> I keep six honest serving-men
> (They taught me all I knew);
> Their names are What and Why and When
> And How and Where and Who.

Start your questions with any of these six words and you will learn. Your questions must go deeper, though. You are searching for people's feelings, what frustrates them, angers them, bores them, or wastes their time.

Questioning without listening is a waste of time. As Stephen Covey says in his *7 Habits of Highly Effective People*, "Seek to understand before you seek to be understood." This means understanding the feelings behind what people are saying.[1] If a person has a lot to say, ask if you can take notes so you won't forget what they said. Above everything, understand their joy and understand their pain.

Another technique to use here, which will be familiar to people in quality management, is the *five whys*, a very simple tool and a great way of getting to the heart of an opportunity. I remember sharing this technique with a group when I was delivering a continuous improvement training course. When I role-played the five whys, one of the guys in the group said, "Just like my kids! Why daddy? Why daddy? Why daddy?" Kids are not afraid to ask questions. Follow their example.

NOTE TAKING AND TECHNOLOGY

If you have a notebook that you are proud of, then you will continue to take notes. Make your notes as cryptic or as lengthy as you wish. This has to be your own style.

The notebook you use can be leather bound, moleskin, or whatever, but let it have style and make it be something that you are proud to own. The paper inside is just as important. When paper is good, your pen flows on the paper. That will inspire you to further thoughts. I say notebook because it is more convenient. You can write faster than typing on a smartphone, and you can also sketch diagrams. This may change as tablets improve. Do not use a voice recorder; they intimidate people. Some people like index cards because they are easy to organize. You can also buy index cards in the form of Post-It pads, which makes them easy to carry. I like Post-It notes with lines on them.

I do take notes on my smartphone and then e-mail them to myself. When I return to my desk, the notes are waiting for me on my PC and they are already typed. The other piece of technology I really like for developing ideas is mind map software (see Figure 13.2). I use Mindjet but there are others for you to try.

I can't stress enough the importance of note taking in the innovation environment. You will often have an idea one moment and forget it the next. Look at the cartoons of Leonardo da Vinci. Those are the notes of a great artist and engineer. Note taking also has the added bonus of "clearing the mind" so other thoughts can then emerge.

Your notes can be verbal or graphical. If you are graphically oriented, the mind mapping technique will probably appeal to you. If you have mind mapping software, you will find it an excellent way of laying out your thoughts.

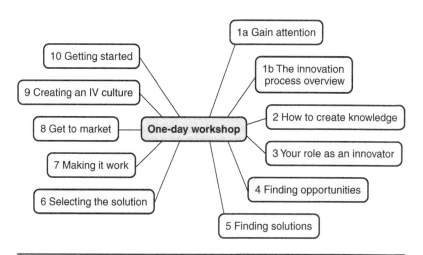

Figure 13.2 Mind map for an innovation process training session.

As you make notes, circle your keywords. Keywords are very important when you return later to read your thoughts. When you are back at your desk, use Google to search for more ideas on the keywords you have circled.

NETWORKING

Networking and ideation are two skills that go hand in hand. I've talked about ideation earlier in this book, and in this chapter I've talked about the skills you should use when you are networking. Let me just focus here on networking attributes. They all center on my good friend "balance":

1. Balance with people who have something to give you and to whom you can give.
2. Initiate the giving, balance the input and output.
3. Balance work and nonwork interaction.
4. Balance your outgoing ideas with quick responses to incoming ideas.
5. Balance trust and respect.

I will talk specifically about networking in the next chapter.

IDEATION AND CREATIVE PROBLEM SOLVING

I talked about ideation in Chapter 5. This is networking on a smaller scale. But remember, with higher social confidence, we create more-radical ideas. Don't expect a rare and occasional ideation session to generate innovation. It has to be practiced every day. The individual brain is not the source of creativity; it works much better interacting with other brains.

Problem solving is familiar territory for people who have had quality management training, but it is traditionally linear in its approach. Nonetheless, people who are left-brain thinkers may want to develop their skills further by pursuing TRIZ techniques, which I mentioned in Chapter 7, "Connecting to the Solution."

Connecting is a key part of problem solving. We are all familiar with the "nine dot" problem, or what is often called "the riddle." For example, many of us have seen this old riddle, which is a good example of thinking outside the box (see sidebar).

> **INNOVATIVE THINKING**
>
> You are driving down the road in your car on a wild, stormy night, when you pass by a bus stop and you see three people waiting for the bus:
>
> 1. An old lady who looks as if she is about to die.
> 2. An old friend who once saved your life.
> 3. The perfect partner you have been dreaming about.
>
> Which one would you choose to offer a ride to, knowing that there could only be one passenger in your car?
> Think before you continue reading.
>
> ---
>
> This is a moral/ethical dilemma that was once actually used as part of a job application. You could pick up the old lady, because she is going to die, and thus you should save her first. Or you could take the old friend because he once saved your life, and this would be the perfect chance to pay him back. However, you may never be able to find your perfect mate again.
> You won't believe the answer . . . .
> Sometimes, we gain more if we are able to give up our stubborn thought limitations. Never forget to "think outside of the box."
> The solution is at the end of the chapter.

BEING PREPARED TO EXPERIMENT AND TO FAIL

Remember WD-40, named after the 39 previous failures? That was a testament to the learning that came from those previous failures. We make mistakes every day, and yet we keep repeating them. House buyers view homes on weekends, not weekdays, and yet when we are selling a house we expect a buyer every day. When we travel we take too many clothes, and yet most of us keep doing it. Those are just a couple of mistakes that many of us repeat, and they are usually in activities we don't do very often. We are trained not to make mistakes, so when they happen we walk away. Never forget the lessons learned in any stage in any project. They do two things: they create knowledge for the next step, and they make us feel better after a bad experience

GIVE YOURSELF THINKING SPACE

A couple of years ago, along with a couple of friends, I visited the NASA site at the Kennedy Space Center in Florida. At the end of the visit—and it was an amazing experience that I do recommend to you—we visited the gift shop. I browsed through the memorabilia and came across some bumper stickers. They had the NASA logo and simply said "I Need My Space." I bought two, one for my wife and one for myself.

"Finding space" is about "time and place." My own "place" is anywhere that gives me space. In the summer, I live on my deck. When I travel, I think and write in coffee shops, bars, and restaurants. My "time" for "notes" is any time, but for heavy-duty writing, it is in the morning. I recall watching an excellent TV series on the history of mathematics. Episode 3 was about the French mathematicians of the eighteenth century. Apparently, Descartes did all his best work in the morning in bed! Maybe it's better to stay in bed than go to work! Some of the world's greatest books have been written in jail!

Most writers do not leave home without a notebook. Keep a notebook in your car or in your travel bag. I've already talked about the importance of note taking in a notebook you like that works for you. My good friend Nora Camps gave me my first "ideas book." Nora gave me an interesting tip at the time. She said, "Once you've written down your thoughts, close the book, and then at the end of the month read through your 'bright ideas.'" Find a book you like that works for you.

Either way, when you find space, allow your thoughts to wander, and transfer your tacit thoughts into explicit thoughts.

THE GIFT OF TIME

Being able to give time to yourself is definitely a skill for some of us. Whether it is taking a bath, watching the sunset, or taking a walk, time and space are something that many of us do not allow ourselves. If you have the gift of being able to give yourself time, then move on to the next page. If not, let me tell you some of what I have learned. When I lived in Europe, people did this well. The culture allowed and encouraged "time giving." When I moved to North America, I found that the skill was eroded. The British and the Germans are inveterate walkers. The French and the Italians are café talkers. Either way, it is time out.

The British sport of cricket takes between three and five days to complete one game. Imagine that in North America! If you truly want to discover time and space, go to Africa. An African sunset is one of the world's great experiences.

You have heard the expression "time to heal." Time does that but it also releases those magical inner thoughts and experiences you have accumulated. My first book was dedicated to my parents and also to my two daughters with the words "to Rachel and Sarah, who remind me to stop and smell the roses." That is a constant reminder to me and we should all do it.

My favorite quote on time is another from Albert Einstein. Those of you with a physics background will appreciate it. He said, "The only reason for time is so that everything doesn't happen at once." This is certainly true in the world of innovation, where everything takes time.

STORYTELLING

This has become fashionable in business, and will become vital when a product is launched. In the world of clothing fashions we would refer to a new line of clothing as a "story." The story recognizes that emotion must be uncorked and released for a new product to succeed. There are many books out there on storytelling. It used to be an art. It is now becoming a science.

If you try to convey a conceptual message, most people have to work to integrate the message into the existing body of knowledge in their mind. The mind is looking for the right home for the new concept.

Stories, on the other hand, relate to past experience, they focus on the main point that matters, and people instinctively place the new message into their existing body of knowledge. People are more receptive to stories. They like to be entertained, and stories are more memorable. All of the prophets used parables.

In the world of innovation, wisdom is not conventional. Stories are a great way of communicating the unconventional.

There are distinct stages to a story, whether it is a book or an anecdote, whether it is Homer, Shakespeare, or Hemingway:

Act I (1) The Status Quo, (2) The Challenge, (3) The Refusal

Act II (1) The Test, (2) The Crisis

Act III (1) The Resurrection, (2) The Climax, (3) The Return

This is the structure of the classic three-act play, but even the simple anecdote follows the essence of this structure. There are also distinct characters in the story. The Hero, the Mentor, and the Villain actually may embody themselves into more than one character.

If you have the job of developing or delivering a new product or service, I encourage you to develop the art and science of storytelling. When we sold the humble towel at Christy, we told the story of William Miller Christy, the Sultan of Turkey, and Queen Victoria. People were suddenly buying a piece of history. A simpler technique I often use is Monroe's motivational sequence. It follows these steps:

1. Gain attention
2. Visualize need
3. Answer need
4. Visualize result
5. Call to action

OTHER SKILLS TO CONSIDER

Risk analysis is widely understood in business today, and is often a subset of project management. However, a basic understanding of risk is useful for the connectors as they evaluate alternative solutions.

Project management skills are definitely required by the developers, and if your business does not have these, it is a must to be addressed.

Strategic selling, as opposed to product selling, and development of the value proposition are skill sets the sales people will require. I talked about this in Chapter 10.

The skill of exploration for business partners often resides in the purchasing function and becomes necessary as an organization moves to open market innovation.

Remember, adults learn by doing. As you deploy your innovation plan, you integrate the training, experience, and skill development with the deployment plan.

One last point. There is an enormous benefit from being known as an innovative company. You can attract and recruit the very smartest people because they want to hang out with other smart people.

As I mentioned earlier in this chapter, I am next going to talk more about networking.

"BUS STOP" ANSWER

The candidate who was hired (out of 200 applicants) had no trouble coming up with their answer.

The answer:

- "I would give the car keys to my old friend.
- I would let him take the lady to the hospital.
- I would stay behind and wait for the bus with the partner of my dreams."

KEY POINTS

- Leadership must fully understand innovation in order to implement an effective innovation strategy.
- A one-day workshop will introduce the management and also the change team and project teams to the essentials of innovation.
- Workshops should include practical exercises that relate the learning to the participants' organization.
- The innovation training plan must be synchronized with the innovation plan.
- The early stages of the innovation cycle require the competencies of exploration, observation, interaction, questioning, and listening.
- Note taking is a fundamental skill throughout the innovation process, and technology can help us here.
- Experimentation and willingness to fail are competencies that can be developed.
- Throughout the innovation process we must be able to give ourselves the gift of time and space.
- The skills of problem solving, project management, and strategic selling, which are required in innovation, are generally well known tools.
- Ideation is often taken for granted, yet must be practiced regularly.

14

Networking

It isn't what you know; it's who you know.

—Anonymous

I have talked about the technical and process side of networking, but what about the human side?

All through my twenties, in my early years out of university, I remember those parties where we would sit into the early hours of the morning and try to figure out how to change the world. My friends at the time were mostly chemical engineers, like myself, who had joined the research division of a large corporation. I'm not sure whether it was frustration with the corporation or just something we all wanted to do at that age, but all of us said, "Some time I want to run my own business." We were all looking for that niche or "magic opening" that would provide the opportunity. If you have passed the 30-plus barrier you will remember what a traumatic point in life that is. You truly wonder where your life is heading.

I remember that 30-plus point in my life; it was a point in my time with Courtaulds when I was very unhappy with my career path. I had moved from the research division to the textile division, and the culture was entirely different. The paradox for me was that I enjoyed textiles far more than chemicals. It was more human and far more artistic, and yet at Courtaulds the particular area of textiles I then worked in lacked the trust of my previous environment in chemical engineering.

I made the tough decision to leave the big corporation and start my own business. I decided I could not face myself if I had not at least been through the experience. My parents were horrified. I read a book on starting your own business that included an assessment of whether you were suited to be an entrepreneur. The biggest psychological obstacle to being successful in your own business was if neither of your parents had been self-employed. The odds of succeeding did not look good.

I took a look at what I had going for me. I had learned a lot about textiles and had a good sense of color. I was currently selling fabric to the fashion trade, and I was a very good business planner. The major negative factor was that I had no money.

I decided to set up a business designing and manufacturing women's fashion, an industry that required little capital to enter. I left Courtaulds, cashed in my pension fund, and used half the funds to buy an old car and the other half to buy fabric.

I joked with my friends that I decided what business to go into by asking myself what I liked. I liked clothes and I liked women, so I decided to design women's clothes.

NETWORKS OF FRIENDS

If you don't have family support, you turn to friends. I had two networks: my original group of chemical engineering friends and a new network of friends at a rugby club I had joined since moving house.

Surprisingly, both networks proved incredibly valuable as I developed my business. You would wonder what a bunch of chemical engineers and rugby players knew about women's fashion. In truth, they provided a mountain of opportunities and contacts. The wife of a "friend of a friend" at the rugby club became my design partner. My chemical engineering friends were incredibly curious and kept asking about my new business. Whether you are on your own or in a big corporation, your networks are vital.

My networks have changed over time. I no longer play rugby but belong to a dinner club called The Thunderers, to which I was introduced by another friend, Ian Oliver. I no longer live in the UK where my chemical engineering friends are, but we still stay in touch. I have two other major networks: the professional society to which I belong, ASQ, and the global network of ISO, which gives me friends across the world.

Having friends "outside the box" is essential for you to have a fresh perspective on the world. Having an environment outside your own where you can go and experience fresh messages, relax, and allow your mind to unwind is essential.

If you want to join the world of the innovator, you need to be mixing in unfamiliar territory. What are your networks?

- People in the neighborhood

- Parents from your kids' school
- A professional organization
- A sports club
- A hobby club

Facebook is not the type of network I am talking about. It does not give you that essential human interaction. I do use LinkedIn, which provides excellent points of contact when I want to share general knowledge, and I am on Twitter @petermerrill, which is good for quick contact.

It is only when you are experiencing all the channels of communication—written, spoken, body language and signals—that your instincts will work to their full. Only when that happens will you release the subconscious ideas that are deep within you. That is why face-to-face networking is vital.

List the 10 people you know best and ask yourself which of them:

- Has a different profession
- Has a different country of origin
- Speaks a different native language

Whatever stage you contribute to in the innovation cycle, you need to be networking. Networking is fast becoming the "new best friend" of everyone in business. It has reached the magic point in its life where it is on the cusp of moving from art to science. What that means is we are finally understanding networking.

SUCCESSFUL PEOPLE WILL NETWORK

In the world of innovation, networking has become fundamental, whether you call it collaboration, open market innovation, or communities of innovation; it is woven through the whole practice and body of knowledge.

Ironically, it was the scientists, or rather the mathematicians, who kept it from us for over 200 years. Euler first developed network theory in 1780, and in Chapter 2 I explained the work of Albert-László Barabási, who brought the science down to earth. At the same time as this mathematical approach flourishes, empirical research is converging with the mathematics to give us a real understanding of how this works.

If you look at the success of some of the great artists in history, it was due to their benefactor, who in turn had a powerful network. Amadeus

Mozart, David Hockney, Samuel Pepys—a musician, an artist, and a writer who, although very talented, all succeeded largely through the network of their benefactors and so pushed out their competition. Yes, the ones I mentioned had talent, but there were many others competing with equal talent, and of course, "success breeds success."

Another famous story of networking is in Gladwell's *Tipping Point* in which he compares Paul Revere with William Dawes. Paul Revere "knew the right people" and quickly raised a militia, whereas William Dawes, who set off at the same time with the same mission, disappeared into history.

Linus Pauling, the twice Nobel Prize winner, with his famous saying "the best way to have a good idea is to have a lot of ideas," openly admits that his success is not due to greater intellect but due to greater networking.

FAMOUS COLLABORATORS

C. S. Lewis first met J. R. R Tolkien at Oxford University. They met regularly and had a common bond in being Christians. However, their perspectives were different. Tolkien argued that the idea of a God made things easier to understand, while C. S. Lewis was interested in exploring the "myth" of Christianity. When new ideas developed, they would write. The outcomes were for Tolkien *The Hobbit* and *The Lord of the Rings* and for Lewis *The Chronicles of Narnia*. You can see the similarity in the works of Tolkien and Lewis, but they were not afraid to share ideas. Their success depended on their personal skill and ability.

DIFFERENT TYPES OF NETWORKS

Networks are about sharing knowledge but also about making things happen. If you look at the different stages of the innovation process, you see that the emphasis on these two activities changes.

The first two stages of the innovation process are about sharing knowledge. To find opportunities and conceptual solutions, we need a diverse, dispersed network. People who may not yet be customers, people who may not yet be business partners, and definitely people in other disciplines. Remember, the breakthroughs occur at the intersection of different bodies of knowledge. That "spark of ingenuity" (see Figure 14.1).

What we have learned about this type of network is that having a large number of people doesn't necessarily mean a good network, and that a

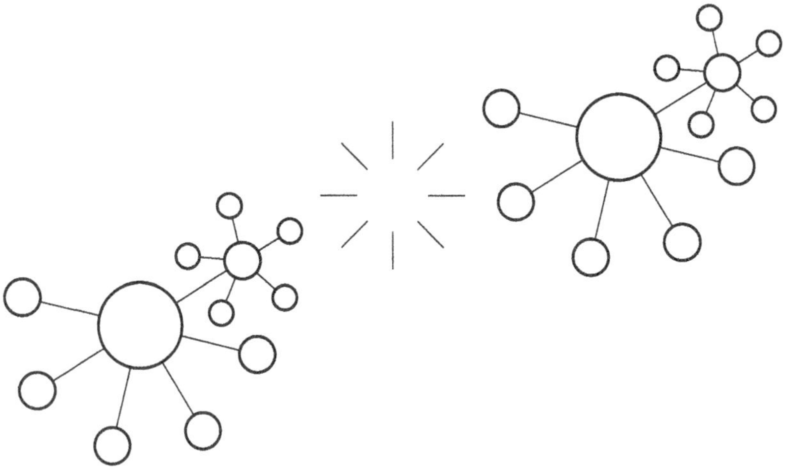

Figure 14.1 The spark of ingenuity.

sparser, well-distributed network is best in these early stages. The really interesting thing we are learning is that indirect contacts or "two degrees of separation" produce the most valuable knowledge.

In practical terms, it has been shown that people who are job hunting are far more likely to find a job through a "friend of a friend" than through any other source, whether direct acquaintances or job ads.

This open but private network has great value when compared to the Internet. It is your private and personal network, and so ideas are contained and captured. You can handle intellectual property far more easily when you are the "hub" of this type of private network. On the other hand, if you search the Internet for solutions, then it is certain that someone else will have that solution as well. In fact, many others will have that solution.

THE "MIRROR" TRAP

Unfortunately, there are many things that militate against us creating this type of open network. People tend to gravitate toward people with similar backgrounds. When this happens, people simply reinforce existing beliefs. They just mirror each other's thoughts.

Next time you go to a party, watch people throughout the evening cluster in their familiar groups. As the evening progresses, people go to the bar and chat with the bartender, who in fact talks with more people than anyone else in the room. At a cocktail party the bartender is the hub and usually gains more knowledge than anyone else in the room.

The "cluster" network has its place in the world of the innovator, but in the later development stage of innovation. These are going to be "like minded" people. The delivery stage will engage the early adopters plus those who are "potential" users, and this will again be a cluster. The cluster type of network, which is similar to your immediate group of friends, will normally have a very high level of trust and collaboration.[1]

Most personal networks are this cluster type, and so if you are seeking to acquire the new knowledge needed by the creator and connector, your network must change (see Figure 14.2).

We generally lack diversity in our personal network. We repeat this mistake in our organization with our recruitment policy, and engage new people who are similar to ourselves. For successful innovation, we need diversity of experience. Let me compare the immediate networks of a couple of people. The first is Hugo, a software engineer in New York, who is from Eastern Europe and enjoys travel. His circle of friends meets regularly at parties, and looks like Figure 14.3.

You can see that only Charlene and Mary have a number of friends outside the circle, and Tom is not even well connected inside the circle. You would think New York would provide lots of networking, but "turning

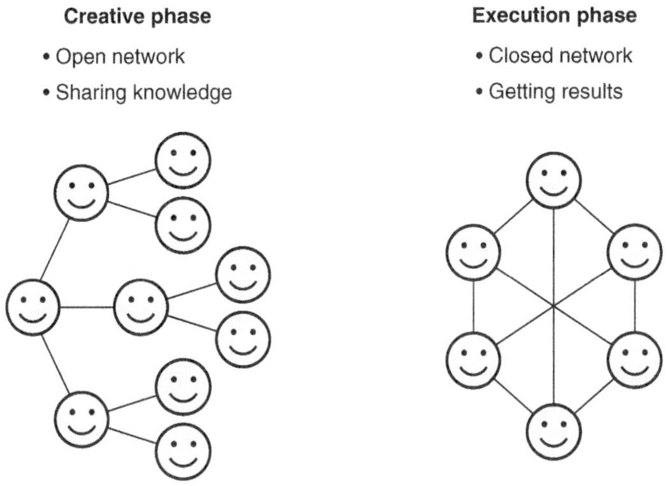

Creative phase
- Open network
- Sharing knowledge

Execution phase
- Closed network
- Getting results

Figure 14.2 Open and closed networks.

	Job	Hobby	Ethnicity	# Friends
Tom	Phone store	TV	New York	3
Charlene	Fashion store	Movies	S. Carolina	10
Dick	Mechanic	Cars	California	8
Pauline	Software	Clothes	UK	6
Harry	Software	TV	UK	6
Mary	Sports store	Skiing	Canada	10

Figure 14.3 Hugo's circle of friends.

	Job	Hobby	Ethnicity	# Friends
Margarite (1)	Housewife	Cycling	Poland	15
Margarite (2)	Mathematician	Chess	New Jersey	10
Renato	Singer	Writing	Argentina	8
Jean	College prof.	Dining out	France	20
Mario	Tour guide	Cooking	Italy	18
Hilary	Tennis coach	Reading	UK	20

Figure 14.4 Carol's circle of friends.

inward" often happens in the big city. Compare this situation with that of Carol, a housewife in Houston with two young children. You would expect her to have a limited opportunity to network, but she has kept in touch with college friends and met others on her travels (see Figure 14.4).

You can see the diversity of the ethnicity in Carol's group, but also the diversity of activities. She is going to learn a lot from her friends and connect with other people through her friends, who all engage in outgoing activities.

So, how do you develop from the scenario where everyone knows everyone else to the one that is diverse and knowledge-sharing? First, let me stress, you don't give up your group of friends! They are important as a key component in your life. However, a "knowledge group" can coexist very happily with them.

YOUR PERSONAL KNOWLEDGE NETWORK

A business network that is diverse and still manageable has 25 to 35 people. Why that number? Well, this is manageable based on experience, and given that I am going to tell you to contact one of the group each business day, this number also means that the "cycle" is not one month but advances by about 10 days each month.

Start by listing your immediate circle (See Figure 14.5).

Now list 35 business contacts that you have. Then go through the exercise of identifying jobs, hobbies, ethnicity, and, if you have the knowledge, show whether they are a high, medium, or low connector.

Now look at your list and see who the closely matched people are. Decide on who to retain. Your list will probably reduce to 25. Now ask yourself, if you brought your 25 to a party at your house, would they mingle happily or would you be concerned about conflict?

If they would mingle happily, your group is not yet sufficiently diverse. Not to worry. That will come with time. You're going to be looking for new members in the months ahead.

Let me remind you, networks are about sharing skills and knowledge and sending messages. They are also about overcoming "stovepipes" and barriers to groupings of knowledge and behavior. They are a great way of accessing skills and knowledge you don't have, and a great way of sending messages that are important. Your messages may appear to be "public," but when compared to the Internet they are very private.

Let me remind you of one other thing. This is a two-way street, and you have to start the traffic flowing. I recall attending so-called "networking"

	Job	Hobby	Ethnicity	# Friends

Figure 14.5 Your own circle of friends.

sessions at the Board of Trade where everyone was "selling" and nobody was "buying." What a waste of time.

You start the traffic by giving. You need to give something interesting. It may be business-related or personal. It will be something you will e-mail, and so is likely to be information that a person would value. It may well be different for different people.

Later, you may move to an event for your network. I periodically hold a breakfast meeting called the Breakfast of Innovators. It is low-key, and there are always a couple of speakers who have an interesting story to tell.

How big can your network grow? Robin Dunbar did some research some years ago and found an upper limit of 150. The danger above this number is that you lose the "magic touch." Maybe technology has increased that number, but if you have forgotten who people are when they contact you, then question your network size.

You are starting to see that in the world of the innovator, relationships are a little different. By that same token, leading a group of innovators is a little different. Let's talk about *leading*.

KEY POINTS

- Networks must provide human interaction to be effective.
- Networking is moving from being an art to being a science.
- Famous people in history attribute a large measure of their success to their network as well as their personal ability.
- Networks need to be diverse and dispersed for knowledge creation.
- Breakthroughs occur at the intersection of different bodies of knowledge.
- Your private network is more likely to contain knowledge that is exclusive to yourself.
- The Internet contains public knowledge, and that knowledge is often not validated.
- Our private networks will tend to include people who "mirror" us, and so will lack diversity.
- To develop networking skill, you should cultivate a personal network of diverse people.
- You initiate your network by sending "gifts of value" to the people in your network.

15
Leading Innovation

A leader is best when people barely know he exists,
when his work is done, his aim fulfilled, they will say:
we did it ourselves.

—Lao Tzu

Some people seek leadership, others have leadership thrust upon them. I seem to have fallen into the latter category. My first experience of leadership was when my rugby club, Stockport RUFC, asked me to captain one of its teams. I guess I discovered two of the essentials of leadership: you must have a passion for the activity, which I certainly had for the game of rugby, and you must understand what the team or the organization needs or lacks in order to succeed. In the case of the team I captained, it was simply a matter of ensuring that the team was not short of players. Anyone who has coached or captained a team will know exactly what I mean by that. Ironically, I was far from being the most skilled player on the team, but I did have the skill of "reading the game," which is vital when you are in the turmoil of a game of rugby. In short, I learned to give the team what they need to do their job

THE ROLE OF THE INNOVATION LEADER

The *innovation leader* has the willingness to explore and take a team into the unknown. This means building trust in the team, and now I will shock you: it does not mean having a vision. I recall at the start of my project to write ISO 10018 on people involvement, one of the team members asked me what my "vision" for the finished standard would be. This was a radical new standard. I said I had no idea and that the finished standard would depend on the team, not me. I had a clear objective, but that is different. That objective was that I was tasked by ISO with developing a standard that would enable organizations to involve their people more closely in the operation of a QMS based on ISO 9001. As time went by, the product evolved, and the final product was designed to meet the needs of the end user. A clear objective or challenge is a critical requirement before you and your team start exploring.

The common idea of the business leader is as a visionary. Traditional leadership style works when a solution is known, but if we venture into the unknown, no one knows what that solution will be. The innovation leader does not set a vision for others to follow. The job of the innovation leader is to create an innovative organization. Leaders draw out individual ideas and assemble them into collective solutions. The approach is to set the stage for innovation. Innovation leaders create the context in which innovation can occur.

Although innovation is often fun, it is also very draining and painful. Collaboration involves disagreement, and this creates stress. Ignoring differences reduces the free flow of ideas and the discussion that innovation thrives on. Leaders must manage this tension so people are willing to share their ideas, but be confrontational enough to improve ideas and encourage new ideas. Innovation requires trial and error, and groups must find their way forward. Solutions arrive that are different from those anticipated.

The leader allows task groups with different approaches to quickly pursue numerous experiments, learn from results, and then adjust the plan. However, its easy to make a decision when something totally failed or succeeded; the real world is "shades of gray."

The leader sets targets, not vision, and by creating a sense of purpose makes people willing to take the risks and do the hard work of innovation. That way, we keep conflict focused on ideas rather than personalities.

A leader uses the theater approach of "improv," avoiding "either–or" thinking and encouraging additive "both–and" thinking. Innovation requires joining ideas together and combining options. A leader must develop "bigger thinking."

The leader develops trust and respect, reduces conflict, and allows everyone to influence. Then everyone has a voice, and even the inexperienced influence decisions. We generate ideas through debate, then test quickly and adjust or "pivot." This also means developing the behaviors of innovation.

DEVELOP THE CULTURE

This is the hardest of all leadership tasks. The looseness required in the creative end of the process and the tightness at the delivery end are in direct conflict, and yet you need both of these modes to exist in the same organization.

Culture is about behaviors, and new behaviors have to be developed. Behaviors always gravitate to the "normal" behaviors of the organization, which are intuitively designed to preserve the status quo. They are designed to avoid wasted time and to increase efficiency. An organization thinking about the future allows itself to think and try out new ideas, and allows itself to fail.

Allowing failure is a tough job for many leaders. Slogans like "Winning isn't everything, it's the only thing" create a fear of failure that in turn creates a self-fulfilling prophecy of failure. A leader must identify the organization's existing strengths and those aspects of the organization that require more work. The aspects that require evaluation are:

- Leadership
- People
- Processes
- Knowledge
- Culture
- Results

This is detailed more fully in the next chapter in the *organizational assessment*. Following the organizational assessment, you need to form the core group and educate both this group and the management team in the essentials of innovation. You then start to develop more communities of practice on your road to an innovative culture. In order to build critical mass, you need a short-term win while at the same time developing the vision of innovation. An early win is critical if you are going to make people believe in your desire to innovate.

As a leader, your immediate actions need to be:

- An organizational assessment
- A management workshop

- Create a core team
- Secure a short-term win

LEADING CREATIVITY

I was keynoting at an innovation conference in Mumbai, and as I was leaving the hotel that evening I was approached by a person who was in the audience. He was a general manager of a large steel fabrication plant nearby. He told me that he was interested in my speech on innovation and had noted that I was also involved in ISO. He went on to say, "Since my organization has introduced ISO 9001, there has been a noticeable loss of new ideas and creativity."

The following day, he showed me around his plant and we looked at the documentation of his QMS. It was not hard to see why the problem had arisen. I explained the cause of the problem earlier, and I describe the solution at the end of the next chapter of this book.

However, the point of my story comes just four weeks later, when I spoke on innovation process and culture at the ASQ QMD Conference in New Orleans. After my talk, I was approached by a member of the US Navy who was responsible for shipyard activity. He said, "Since my organization has introduced ISO 9001 with our subcontractors there has been a noticeable loss of new ideas and creativity." This was chilling: identical statements four weeks apart on opposite sides of the planet from two people who had never met.

Leaders of organizations frequently express concern about the loss of creativity from implementing their quality management system. As a member of TC 176, I hear this complaint regularly. The improvement component of the standard is not bringing the "improvement" that leaders desire. Deming indicated that 80% to 90% of the problems in a process are the result of the system in which the process operates.

In one of the IBM Global CEO surveys, CEOs said threats and opportunities were coming faster and with less predictability than at any time in the past. These threats and opportunities are intertwining to create increasing complexity, which is the biggest challenge facing their organizations. They then identify "creativity" as the single most important leadership competency to address this challenge of complexity. However, they also see the need to equip their entire organization to be creative rather than isolate "creative people" in departments like R&D. Employees at all levels must question "the way we've always done things." Enabling creativity and challenging the status quo is the first step on the road to innovation. This is the business leader's prime task.

LEADING THE PROCESS AND THE PROJECT

Every leader has a contribution they can make to an innovation process and project. Look again at the self-assessment in Chapter 3 for an indication of that role, and then where you can actually contribute to the process. My own score is unusual. I am a developer, which comes from my engineering background, but a very close second is my creator score, which is the artist in me. What I have found from empirical experience is that I made my own best contribution by acting as a link between sales and design people. In the Christy business we had people who were much better than me at both of these functions. However, as in many organizations, the business structure did not lend itself to good communication between sales and design. A leader must always be looking for the primary needs of their organization. This way, the leader gives the people directly what they want, and so strengthens their commitment to innovation.

LEADING THE PROCESS

A leader's first task is to help people understand the innovation process, which I discussed in Part II. Most business writing and popular perception totally misunderstand the process. Business strategists look at the simple model of market development, product development, and product delivery presented in Chapter 1. This model needs more detail in order for the leader to manage it. These additional functions are shown in the model in Figure 15.1.

Popular perception goes wrong when it groups activities into sales and marketing, design and development, and operations. The groupings should be (1) marketing with design, (2) development, and then (3) operations with sales.

Every leader has two innovation roles in their organization. One inside the process as a participant and the second outside the process to ensure that it is functioning. I already talked about involving people. A leader must get their feet wet and their hands dirty as they change to an innovative culture.

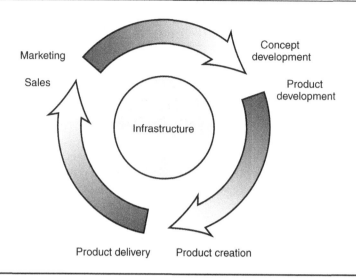

Figure 15.1 More detailed process model—showing feedback.

Having agreed to the challenge, a leader must be close to the team to ensure that all members are aligned with the challenge, and also to draw on ideas as the journey unfolds.

The leader's job then becomes more complex: staying connected to the end goal but also staying connected to the parent organization. This means having a clear process that is understood by the team and that is used regularly as a check and balance for the mission. I have described the innovation process previously in simple terms, and that was to develop understanding. The reality is, of course, more complex, and Figure 15.2 emphasizes the iterative component of the innovation process.

The first step in the process is to find the opportunity, or unmet customer need, and the second step is to find conceptual solutions. In the real world, we do not crash on and select our own preferred solution. We test the concepts with the customer. For a manufactured product, 3D printing has become a huge asset here.

The internal looping in the figure shows how we constantly test, and a critical job for the leader is to ensure this happens. The leader must stay close to the team and close to the customer. You should expect to have a clear identification of customer *needs*—and I stress, *not* the *solution*—in the first couple of months of the project. The concepts you place in front of

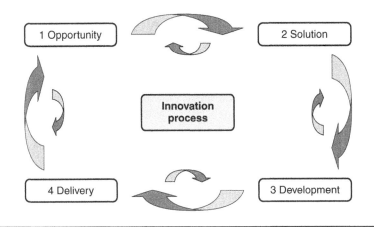

Figure 15.2 More detailed process model.

the customer are to stimulate their thinking and demonstrate the "possible." In a past life, I worked in the world of fashion. A fashion designer will create a "collection," and it is exactly that. They collect their ideas, and among the ideas will be radical new ideas, which they are testing with their potential customer. This is especially true with couture. This way you test and learn is through a continuous "what if" process to draw out alternatives. It is especially important for the leader to draw in the unhappy customer during this early stage of the process.

As ideas develop, the leader must challenge the team. The advantage a leader has is a degree of objectivity, and this enables them to challenge assumptions. At the second stage of conceptual solutions or ideation there will be a lot of assumptions. The leader must ensure they are collected and challenged. Every assumption must be tested to find (1) what you don't know, and (2) what other choices you have.

This again must engage the customer, and will have the advantage of strengthening the customer relationship.

As assumptions are tested, the associated risk will be evaluated, and this is where the leader must connect back to the parent organization and ensure that the leadership of the parent fully understands what is happening. They may even have the job of ensuring that the leadership understands the innovation process. This may well be a half-day or one-day in-depth session even though business leaders will say they can "get it" in 40 minutes. As the solution evolves, one more issue the leader must address is the business model for delivering the solution.

SET THE CHALLENGE, NOT THE VISION

Amazon set the challenge of delivering its products more rapidly, and hence its pursuit of the use of drones. In the '60s, JFK set the national challenge of a man on the Moon and safely returning to Earth before the end of that decade.

The next role of the leader is to agree to challenges, involve the people, and give them resources. Challenges come from market trends and opportunities, and knowing your organization's core competencies. When I say trends, I mean mega trends like technology, environment, and terrorism. These are, of course, subsets of the biggest trend of all, which is globalization. This broad brief to the market researchers should then leave them "loose" to find opportunities.

The leader must set challenges at the marketing stage, knowing the market strengths, such as brand and operational strengths, of your organization, as well as your creative strengths, where you should be focusing for your new opportunities.

This challenge must not be narrow or you will miss opportunities; Apple would not have gone into the music business if their leadership had thought, "We make computers." On the other hand, the challenge must be one that uses your competencies or you will spend forever finding new resources. Apple, again, keeps moving the goalposts so competitors can't catch them.

This is one of the toughest jobs for the leader. Jim Collins in *Good to Great* defined it as the *hedgehog principle*, expanding on the quote from Erasmus:[1] The fox keeps jumping from one opportunity to another. The hedgehog rolls into a ball, focuses, and prevents others from attacking.

RESOURCE THE PEOPLE

Giving time is something most leaders do very badly. The obsession with short-term efficiency is the greatest inhibitor of innovation. People are not going to innovate when they are exhausted in the evening, or at lunchtime when they are hungry.

Innovation is a prime-time activity. Prime time is in the morning, when of course everyone spends their precious time in routine meetings.

The most overlooked business resource is *knowledge*. Knowledge comes from education, and education is probably the item that most leaders will cut in a tight budget.

Providing resources means ensuring that you have the right people at each stage of the innovation process. Assess your process (see Chapters 6 through 10) and have your people self-assess. Ask where you are deficient. You may have the right people component, but your process may be weak. For example, you may have a strong presence of connectors at the problem-solving stage, but your process is weak through lack of a defined process or unclear objectives. The leader must ensure a balance of people and process at each stage.

Resourcing also means providing the right tools and technology needed. Be careful here. It is more about developing IT skills and competencies, and not about huge investment in the "cool tools" of IT.

ENTREPRENEURSHIP

There is a lot of interest in entrepreneurship, and this ties to innovation. Universities offer combined courses in innovation and entrepreneurship. However, the two activities are definitely different. They are complementary, but they are different. The key facets of the innovator are focus, creativity, and teamwork. Contrary to popular image, neither Edison nor Jobs were innovators. They were far more entrepreneurs. Edison did not invent the light bulb, and Jobs did not invent the MP3. They both saw a great idea and built the infrastructure within which these ideas would have value.

I have always been an entrepreneur at heart, from when I ran my first business and at one time ran a market stall. I have been through the "school of hard knocks." Today, I find myself to be an "entrepreneur enabler." All entrepreneurs are driven, and I have discovered that my own "driver" as an entrepreneur is my desire for independence.

The opportunity seen by the entrepreneur is more often financial, whereas for the innovator, it is about a "new way of doing something." Creative people are often very poor at the commercialization of their idea. They are passionate about their idea and will often give up everything for their idea. They often do not manage money well, and will "sell their soul to the devil" in order to see their idea reach fruition.

In order to succeed, an innovator has at some point to become an entrepreneur, or partner with an entrepreneur. Unfortunately, the entrepreneur often then becomes the leader as a result of their financial skills. We have seen this lead to business demise when the entrepreneur and the innovator have different objectives. This was shown with the split of Steve Jobs and Steve Wozniak.

Good entrepreneurs are risk takers, but they are also good money managers. They calculate risk and act accordingly. They are frequently not committed to "the idea," and when a business reaches a certain size, they will sell it for financial gain.

The entrepreneur is a self-promoter, and this is one necessary attribute for an entrepreneur, but not the only one. This must be tempered with the recognition of and collaboration with colleagues. We know that strong collaboration is fundamental to innovative success.

No individual will have all the attributes to be both an entrepreneur and an innovator. This is why I come back to the importance of collaboration between those who are primarily innovators and those who are primarily entrepreneurs.

MAKE THE DECISIONS

Creating a culture of freedom and choice allows ideas and opportunities to emerge. The tough choice for the leader with lots of ideas is, which ones to run with?

It is so easy to choose based on short-term return. Most businesses look for three years' ROI. I mentioned already that Xerox has found that some of their best new products can take seven years to get the ROI they need. Risk must be calculated, then managed. Risk must not be avoided.

The leader needs nonfinancial metrics at each stage of the process. A leader with a strong financial background may have the toughest job here. Measuring "exploration" is one of the first challenges. The creators are looking for opportunities. At the outset, the metric may be the number of customer interviews, moving to the number of opportunities later. Measuring the connecting or problem solving means, how many potential solutions are you finding?

The selection stage is rich in measurement opportunities. This is where the innovation process connects to the strategic planning process of the business. The leader is now in more-familiar territory. Measure failure—remember, Intel said if they are not getting a 10:1 failure to success ratio, then they are not taking enough risk. Measure your ratio of long-term to short-term ROI selection (see Figure 8.1). Measure completion time from selection of solution to first sale.

This takes us into the development stage of the innovation cycle. Here the measures are about speed and the more tricky user-friendliness (see Chapter 9). Finally, at the execution stage it is about measuring speed of delivery and measurement of sales growth, again more-familiar measurement territory for the leader. I will talk more about measurement in Chapter 17.

TAKING ACTION

The leader must periodically step back and review the health of the innovation process and the organization as a whole. I dislike the word *review*. It is so imprecise. It literally means "look again" and it does not infer taking action. In actual fact, review means comparing results with expectations and taking action on the shortfalls.

If you were engaging in serious change to an innovative organization, I would recommend a quarterly management review of your innovation process. This is where your innovation team looks at each stage of the innovation process and identifies whether the targets are being met—whether soft targets or hard, whether the process is operating as required, and whether the process is being properly resourced. Your innovation plan is then updated. You may choose to do an innovation test at points in the process where concerns exist. The typical agenda for the innovation review is:

- New opportunities
- New solutions
- Selection of solutions
- Measurement of results
- Action points in the process
- Skill development
- Technology needs

USE A PROJECT TO INITIATE CHANGE

In the early stages, you may be just looking at your innovation process and building the process, so the agenda will be structured around process design and resource development.

This is a different kind of leadership, and it is at the project level. The job of the leader is to take an innovation project from opportunity to execution. They lead from within. Their primary job is to build and sustain the group. Managing the changes in the team's makeup and the team culture is the most demanding task. I will talk more about managing change in Chapter 19, but this is a key issue in an innovation environment. The innovation project team will shift in content as you move through the stages of the process. In the first two stages, you will be in a loose mode, and monitoring activities is critical. There may be people who have a separate agenda, and they may feel strongly about it. They must be allowed to pursue that agenda, but not at the expense of the rest of the team. This is where the art of leadership becomes important. Over-controlling will kill new opportunities.

The leader needs to direct opportunities into the problem-solving stage, and there may be solutions that are discarded but that need to be carefully stored for potential future use. The solutions have to be assessed so that useful information goes forward to the selection stage.

The innovation team presents its options to the strategic planners, and once selection is made, there is radical change. Now we run for the line, and the innovation project leader may even change. The team culture changes from loose to tight, and the team's makeup shifts to a predominance of developers and doers. This is where the traditional skills of project management kick in.

I was one of the team members that wrote ISO 10006:2003, *Quality management in projects*, and I refer you to it as one of the best bodies of knowledge in this area.[2]

Throughout all of this activity, speed matters, and this is the other prime attribute of the innovation leader. They have to ensure that both the leaders of the business and members of the team understand the "need for speed." Someone, somewhere, will have seen the same customer need you have found and will probably be working on the same solution. The leader must maintain communication with their team, the organization, and the customer who will use the new offerering (see Figure 15.3).

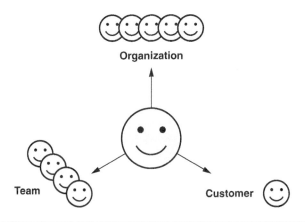

Figure 15.3 Leader's connectivity.

Experience shows that 75% of innovation or start-up projects fail to meet objectives. This is mainly the result of having a grand, inflexible plan at the start and failing to get customer feedback. Fail early and be ready to change direction, or to use the fashionable term, to "pivot."

So, next let's look at the type of organization the innovative leader needs in order to succeed.

KEY POINTS

- A leader needs to know their own "innovation aptitude."
- The traditional functions of an organization are grouped with sales and marketing separate from design and development, and both separate from operations.
- The traditional grouping of functions impedes innovation.
- A leader's first task is to set strategic challenge for innovation.
- Strategic challenge is based on market need and the core competencies of the organization.
- Giving people "creative time" is one of the hardest tasks for a leader.
- A leader must ensure that the right people are involved at each stage of the innovation process.
- A leader makes the tough choices on which ideas to pursue.
- Monitoring progress through nonfinancial measures can be a tough change for leaders who have a financial background.
- A leader must ensure the switch from loose to tight mode at the tipping point.
- The leader must conduct a regular review of innovation projects.
- Above all, the leader must lead the change to an innovative culture by communicating a vision of innovation and recognizing innovative behavior.

Part IV

The Implementation

16

The Innovative Organization

Once an organization loses its spirit of pioneering...
its progress stops.

—Thomas J. Watson Sr.

Gerry Kavanaugh made paddles for dragon boats. Dragon boating is a classic example of a new sport where the people involved in the sport have innovated the equipment they use.

Whether it's mountain bikes, snowboards, or kayaks, the people involved in these sports have seen the need to refine their equipment to make it more effective in its use. This is customer-driven innovation, but you will notice that these are a different kind of customer. They live on the edge. There is a saying, "If you're not living on the edge, you're taking up too much space." These people live for innovation because they have passion for their sport.

Gerry Kavanaugh certainly lives on the edge. Gerry is one of my clients, and he drove in Formula Ford motor racing. While traveling the racing circuit he discovered dragon boating. This sport, which as you would expect comes from Asia, has 22 people in a canoe who race on open water over a distance of 200 to 2000 meters.

Gerry found the sport exciting and got involved. He discovered very quickly that one of the biggest problems in the sport was that given the tremendous force exerted on the paddles, the wooden shaft had a short life, and almost every race experienced "paddle fractures."

Gerry's experience in formula racing had exposed him to carbon fiber, and he connected the two experiences and developed a carbon fiber paddle for dragon boating. His auto racing experience enabled him to develop equipment to manufacture carbon fiber paddles. Gerry's business took off, and he quickly had more business than his facility could handle. He won't mind me saying his workshop became hard to manage. Gerry's was a classic start-up business.

I worked with Gerry to help him grow to the next stage in his business. He developed structure as he grew, but retained that flexibility that has made him so successful to date. Gerry's mantra is "You never stop innovating!" He now makes carbon fiber aircraft components, and his organization has changed again.

A "FLEXIBLE" ORGANIZATION

You may be surprised, but Gerry used the framework of ISO 9001 to get this structure. The standard gives him the framework to grow his business. He has mapped his business flow and identified his primary risk points, and at each of those risk points he is monitoring his processes. He has started a regular performance review of his business, and nonfinancial indicators support the financials. He has identified the key skills needed by his people and developed a competence database so that as he takes new people on board, they have the right competency to do the job.

Start-up organizations are exciting. How many times has the CEO of a large, rapidly growing, and successful business turned to a best buddy who grew with the organization and said, "Remember how it used to be?" This is a yearning for those old days when the business was small and hungry, everyone knew what was happening, and you could turn the organization on a dime.

The creative phase is followed by selection of preferred solutions (see Chapter 8) and then the business model for delivering the solution should also be developed.

GROWING THE ORGANIZATION

Success breeds growth, and in *Do It Right the Second Time* I explained the limits to growth of an organization. You can generally reckon that for good lateral communication your span is about six people (you may recall de Bono's law). In terms of layers, reckon on three layers or two degrees of separation in the organization as the limit of good communication (see Figure 16.1).

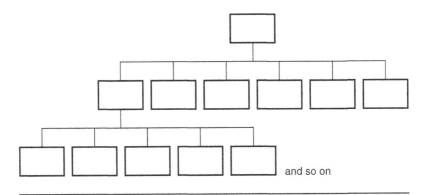

Figure 16.1 As the organization grows, it loses agility.

Beyond these limits, your flow of information, and hence your flow of knowledge, will suffer. What does that mean in terms of the size of a business unit?

Layer one. Six people—that was the core group in the good old days of start-up.

Layer two. This was created when each of the original six got a team working for them, again about four to eight people in each team, and you grew to about 40 persons. Over 90 percent of businesses never grow beyond this. It may be due to a leader who won't let go and keeps the business at this size, or it may be the absence of business controls, and each time there is growth it lapses back to the "40 limit." Another factor for the next growth stage may be that an organization needs a different kind of leadership with more advanced leadership skills and that understands a larger organization. That's why many have to sell their "start-up" to a larger organization.

A few go beyond and become an "organization." You will see that the limit here becomes $6^1 + 6^2 + 6^3$, or about 250 people. The organization provides a successful product or service, and they probably develop a management system that makes things run smoothly.

Therein lies the danger. After three, five, even 10 years of successful operation, the market changes. Some businesses change (painfully) with the market, others do not and they go out of business.

The challenge for the growing business is to identify the new market and product ahead of time, and of course that is what innovation is all about.

However, if you are already a large organization, this presents a different challenge. As you change, avoid brutal restructuring and learn healthy disengagement; avoid downsizing or mass firings, and continually adjust resources.

THE "SPIN-OFF" ORGANIZATION

Most large companies are better at executing what they do today than changing for the future. Beyond the 250 mark, an organization frequently loses its agility and becomes set in its ways. The "elephant can no longer dance." A spin-off organization is often the answer. If you have moved beyond the 250 mark, there is also a chance that your organization has split like an amoeba into what we appropriately call *divisions*, or overcome the problem by creating a spin-off organization to develop and deliver a new offering.

A spin-off innovation business rarely cohabits well with its parent. Letting go, borrowing assets, and learning new behaviors are tough tasks.

Many spin-offs established with a mission of innovation make the mistake of borrowing too much from the parent organization. They borrow the parent's assets and so replicate the old behaviors. The new business does not become a learning organization (see Chapter 2). The spin-off must operate with right-brain creativity in its early stages, but it does need a left-brain structure at later stages to connect it to the main organization because of its need for financial resources.

The secret of success for the start-up or spin-off organization is having a network structure, independence, and diverse people.

There is an apocryphal story from Corning Glass that illustrates the pitfalls associated with creating a spin-off.[1] The lessons can equally be applied in organizations of less than 40 people when you are setting up your first innovation team.

Corning saw advances in biotechnology that spawned an industry dedicated to genomics. They saw a growth opportunity in supplying glass slides for labs, and created a spin-off. They were experienced in supplying to industrial manufacturers, had strong intellectual property rights, and were "world class in glass." However, they would be selling to an unfamiliar customer base that wanted low cost and convenience.

The new business adopted Corning's rigorous five-stage model for new product development. Management positions were filled by Corning insiders.

The behaviors and culture were identical to that of Corning. However, the usual methods for correcting problems did not work in the new market, and they missed deadlines.

After two years, they appointed a new general manager from outside the organization who reported directly to the president, and the organization started to follow a looser innovation process. Within a year they had a success. The previous GM was highly capable but had been asked to operate in a context where success was unlikely.

NEW KNOWLEDGE, NEW BEHAVIOR

There are some key lessons to be learned from this and other similar experiences.

The spin-off organization has "knowledge generation" and not product delivery as a primary task, so it must avoid being hierarchical. This is essential when creativity is required. Even as it grows, it will need to work on retaining its network. New knowledge it requires will often come from outside, and new behaviors are needed, so it must disconnect from internal knowledge sources. This means a separate IT system. You are creating a new "brain." New behaviors mean disconnecting from the existing HR function.

These are very visible actions and make a clear statement of autonomy. The people who are recruited should be different, but they should not all be creative artists. You need to use the assessment in Chapter 3 to create a balance of aptitudes.

One more human resource needs careful thought. The leader. This person is not necessarily a creative genius; they are the person who can draw the creativity out of a genius. Recall my reference to Steve Jobs in Chapter 11. The leader facilitates innovation. They also have the task of getting funding and other resources. The advantage of an outsider is that they do not have the baggage of an insider. However, their disadvantage is that they do not know how to navigate the politics of the existing organization. My experience is that they should report at one level higher, say, direct to the CEO, than might normally be the case. Sadly, one of their tasks will be dealing with tensions, even resentment, if the parent organization has developed a "traditional" culture.

In recruiting an outsider, it will therefore be doubly important to ensure that they "fit" with the core values of the parent. However, one thing to beware of is that the actual behavior seen in the parent organization may not always relate to the original core values of that organization.

Even though the leader has a direct line to the CEO, experience has shown to keep the spin-off hungry for cash. Not so hungry that they are despondent, but hungry enough to create a resourceful culture.

THE BRAND

One aspect of the main organization that must be retained in the spin-off is marketing and brand. The marketing people will be delighted, and yet so often they are overlooked. The spin-off typically just engages R&D folks

What's branding got to do with it?		
Technology	Bell	Intuitive—the phone rings a "bell"
Fashion	Polo	The sport of the rich and famous
Technology	iPod	"I" means "innovation," but also I means me and my personal device
Domestic	Tide	Means clean: clean as the ocean
Leisure	Grey Goose	Ice cold
	Kalashnikov	Violence—brand transfer to vodka
Technology	Quicken	Getting the job done faster
Fashion	Gap	A hole in the wall

Figure 16.2 What's branding got to do with it?

and maybe an occasional production person so they can "keep in touch." Market research people should have the key skills for finding the opportunity, and sales and brand people are needed at the delivery end of the process.

Brand is definitely a key asset that the organization should use if it is the right fit. The Tide brand of Procter & Gamble, although associated with laundry soap, in the mind of the user means "clean." They created "Tide to Go" as a cleaning stick. Crest, although initially associated with toothpaste, transferred brand identity easily to toothbrushes. I led the spin-off of the Christy Brand in the UK from towels to bed linen. The customer thinks "bed and bath" in the same context.

Think carefully about branding; this is where you name your baby. The name must make a connection for the customer. Making the brand name an ego trip for the organization is a waste of opportunity (see Figure 16.2).

KPIs IN A CREATIVE ORGANIZATION

Given that knowledge generation, or learning, is a prime task, the *key performance indicators* (KPIs) in a creative organization need to be nonfinancial and focus on issues like number of market opportunities found, number of potential solutions, speed of resolution of bugs, and speed of client acceptance. The business meeting will focus on these measures and not on profitability. Obviously, cost control is important, mainly from the perspective of using it to finance learning.

For the spin-off, the business meeting needs to be separate from the parent and probably more frequent. The focus will need to be on trends and not results. This means that plans need to be updated frequently, so predictions need to be made on a regular basis. I will talk more about measurement in the next chapter.

STRUCTURING FOR INNOVATION

We do a lot to inhibit this diversity by the way we organize our businesses. We classically team up product research with product development. Those are the people on the inside, and they are similar kinds of people, so we group them together in R&D.

Then, we team market research and market development (or sales) together because they are more outgoing than those R&D folks and they spend most of their time outside the building and need managing differently.

If you think about it, the market research and product research people should be working together because they are dealing with concept, both opportunity and solution. The production and development people should work together because their job is to get the new offering in the hands of the customer.

Another major factor here is culture, or behavior. Remember, as Deal and Kennedy said, culture is "the way we do things around here." The research culture should be open networked and blue sky, with nonfinancial measures. The development culture must be highly focused, closed network, and intent on speed.

The tipping point between these two distinct sets of behavior is where we make our decisions on which products to pursue. Up to now, we have been driven by change. Now we are driving change, and one of the biggest resistors will be "the enemy within."

USING ISO 9001

For information to flow quickly, we need an organization's processes, its people, and its technology to come together as a system. But this system must be flexible. Innovation is seriously inhibited by a lack of organizational agility. ISO/TC 279 Innovation Management was tasked in 2014 with developing an innovation management system, ISO 50500. The structure is the same as ISO 9001 and follows the same high-level structure adopted by ISO for all management system standards. As we move to integrated management systems, the *innovation management system* (IMS) and *quality management system* (QMS) must clearly be linked.

This shows the increasing national concern and also increasing global consensus on how innovation should be managed in an organization. However, there are over a million users of ISO 9001 globally, so it makes a lot of sense to see how we can integrate innovation management with an ISO 9001 quality management system. You will recall, I mentioned using it at the start of this chapter.

People often say innovation and ISO 9001 are a contradiction in terms, but that is only the case if you have used ISO 9001 incorrectly and made your organization rigid instead of agile. We can use the ISO 9001 requirements for innovation management, but we need a deeper level of thinking and application. Systems thinking is imperative for successful change, so systems thinking is essential for successful innovation. I am going to focus on certain key elements of ISO 9001 and show you how to use the activities you conduct in each of those elements to develop innovation management. You will recall, the innovation process follows these stages:

1. Identifying the opportunity
2. Connecting to potential solutions
3. Selecting the preferred solution
4. Developing a working solution
5. Delivering the solution to the customer

SETTING STRATEGY

Clause 4 of ISO 9001:2015 introduces the term "context," and this is where "risk-based thinking" starts. Context means you must ask, "What are the external and internal issues that affect your business strategy?" All of these issues introduce risk and opportunity. The standard is helpful and lists examples of these issues. You should work through them to find which are most important for you and why. Don't forget that you are thinking at a strategic level in clause 4.

The list of examples in the clause starts with the easiest issues. If there has been a change in the legal and regulatory requirements that affects your customer, this immediately creates opportunity *for you*. For the customer, this may appear as a restriction. For you, it is an opportunity to ease the customer's pain *by innovating*. You don't necessarily change your product, but maybe provide a new service to assist the customer.

In the marketplace, change—and therefore your context—is moving faster than ever. Much of the change your customer experiences is the result of technology. Technology eases some problems, but it also creates new needs. Again, find your customer's pain and ask yourself, "How can I serve my customer in this new context?" At the same time, you should conduct competitive analysis and find out what your competition is doing to ease your customer's pain.

The other issues you must examine are the cultural and sociological changes that affect you externally. You are probably tired of hearing about social media, but this ties to technology, and simple matters such as registering online for courses and conferences have changed dramatically.

How many checks have you written in the last 12 months? Can customers pay you easily? Customers also can voice opinions more easily as the world has become more transparent. Even if you provide business-to-business solutions, word on your performance will be more open for people to see. Thus, clause 4.1 introduces *opportunity thinking*, and this is great fuel for the innovator.

Clause 4 also introduces *interested parties* and subtly gets users thinking about social responsibility (SR). Some dismiss this as do-gooding, but increasing evidence shows it should be a vital part of your strategy.

Pepsi has demonstrated this in a major way in its own business strategy, in particular with the work of its CEO Indra Nooyi: "We looked at eating and drinking habits of the people, we realized there was a need to have healthier products and good-for-you products. Then we realized we can not tell the customers that healthier products would cost more and may be tasteless. So, we adopted a strategy and that transformation worked well."

LEADERSHIP, RISKS, AND OPPORTUNITIES

Moving into clause 5, the "input information" for leadership decisions includes the threats and opportunities your business faces. Rita McGrath showed how competitive advantage is transient. This underlines the need for you to evaluate your threats and opportunities continually in this fast-changing world.

Clause 5 on leadership encourages you to start to "set objectives compatible with the strategic direction and context of the organization." The importance of addressing opportunities in clause 4 now becomes evident to the innovator.

Clause 6 covers actions to address the risks and opportunities. This is where it becomes vital not to become risk averse. The internal context in clause 4 guides you to address internal issues, such as values, culture, knowledge, and performance. If you're not achieving the performance you want, you are instinctively becoming cautious, and this is why culture and values become vital.

Innovation culture embraces exploration, collaboration, and experimentation. You must address these in clause 4 as internal issues if you are going to enter clause 6 and break free of risk aversion and into embracing opportunity. Clause 6 also notes that taking risk can also mean pursuing opportunity.

Evaluating Risk

Risk is evaluated from data. Information is gained from data analysis. As we gain more data, our knowledge of risk increases. In the same way that we evaluate threats and opportunities continually, we evaluate risk continually. Here's an idea: when you start to document your ISO 9001:2015 system, don't write "risks and opportunities"; instead write "opportunities and risks." It's a subtle change, but it creates a whole different mind-set. Don't let your auditor tell you that you must write it as per the standard.

Here's another thought: when you write, "achieve continual improvement," there is nothing to stop you from adding the words "and innovation."

Having assessed risk at a strategic level, you must decide what opportunities to pursue. The plan is developed and deployed in clause 6, which asks you to "establish quality objectives." This is where you include your innovation objectives and "retain documented information" on those objectives.

Clause 6 also addresses planning of changes and reminds you that there will be those opposed to change—nowhere more than when we try to innovate. Clause 6 addresses availability of resources, and this becomes one of your critical strategic issues moving forward. Resources are not just budget, but also people and their time. Don't overlook political resources. Senior sponsorship of opportunities becomes essential. The people who oppose pursuing opportunities often will oppose them emotionally rather than logically. Find what they stand to lose, and have your sponsor address that loss.

Planning is not a one-person conceptual activity; it is a vital conversation between business leaders in which you agree on what business opportunities you will pursue. It is also not a once-a-year activity. The plan must be monitored as it is deployed, and your management review is your forum for this. You will now see that management review is covered in clause 9 of ISO 9001:2015.

After developing your strategy, setting your objectives, and creating your plan, you must deploy it properly or your strategy is worthless.

Remember that the key word in systems thinking is "linkage." This is what ISO 9001:2015 will do well for you if you use it well.

The areas of the standard I have addressed here are those we have traditionally neglected in quality management systems. You now have a method to link your QMS to business strategy and to innovation.

WHERE TO BEGIN

The starting point for innovation in the ISO 9001 requirements is customer satisfaction monitoring. You must stop asking questions like "Are you satisfied with our product, delivery, and so on" and start asking questions like "What are you having difficulty getting done, where are you wasting time" and ask for responses relating to both your own products/services and those of your competitors.

It is important to recognize that in tough times it is usually more profitable to find new products or deliver new services to your existing customers than to go sourcing for new customers (see Figure 16.3).

Another source of input to the IMS, and which identifies opportunities, is the analysis of nonconforming product or service. If you're experiencing

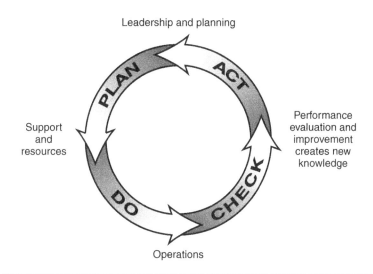

Figure 16.3 Create "new knowledge" from your ISO 9001 management system.

repeated failure here, you have found a clear candidate for innovation. At this early stage in the innovation process, you should also assess the aptitudes of your people and find where they will make their best contribution to the innovation process. This connects to the competence requirement in ISO 9001.

FINDING SOLUTIONS

We then move to the next stage in the innovation management system by using the data analysis requirement in ISO 9001. Data analysis of customer satisfaction and product and service failure gives us the opportunities where we should focus. At this point, we should recall the famous saying attributed to Albert Einstein: "If I had an hour to save the world, I would spend 55 minutes defining the problem and five minutes finding the solution." We need a clear definition of what the customer can't get done or the process failure that leads to product and service requirements not being met. Good data analysis provides this.

Conceptual solutions then come from using *continual improvement*. You should then have a problem-solving methodology such as I discussed in Chapters 7 and 13. It should be a methodology that takes you "out of the box," engages other people inside and outside the organization, and is not something done by one or maybe two folks in the quality department. This is a key element in connecting to an innovative solution.

The best solutions come from accessing collective knowledge. This work is best done by those who can best connect with other environments and industries. For each of these ideas, you also need data on time, cost, and risk for implementation. Risk assessment, which I discussed in Chapter 8, should be external as well as internal, and we often forget this. Potential new suppliers are evaluated. If you have a new product, there is a 90 percent chance you will need a new supplier. In fact, there is a high probability you may decide to have a new product entirely made by a new supplier, or a new service entirely delivered by a new subcontractor. New supplier evaluation is performed by using your internal audit capability. Too often, we evaluate a supplier's product or service and not their QMS. You need to know how well your supplier is managed, and mitigate any risk.

MANAGING THE SYSTEM

Your management review should be conducted at least quarterly. If it is not, then you are not managing your QMS. The management review process is where you select the solutions you will focus on, and it is also where you monitor progress. Selections are made based on good risk data. Management review does not have to be a conventional meeting, but its decisions must be documented.

There is now an overwhelming "need for speed." I have already said that someone, somewhere will be working on the same problem as you. Ironically, design and development (D&D) in ISO 9001 is where too many slow down. It is the largest element of ISO 9001 and aims to force rigor and discipline. I find all too often that the conceptual design work is left until the development stage instead of being done in the earlier problem-solving stage prior to selecting solutions. A well-defined project plan and timeline are needed together with well-defined design review stages. Good design is like good golf. You take your time when you hit the ball (the design review), but you move quickly between shots. The design review output is also fed back to management review in case resource issues need to be addressed.

BENEFITS, NOT FEATURES

The game changes again at the final stage. The focus on product and service features in the design stages means we can often forget the original reason for starting the innovation. This "reason" for the innovation is the "benefit" to the customer in addressing their original pain. Develop a *value proposition*, which describes the benefits to the customer that your new offering will provide and, importantly, address the customer's pain. ISO 9001 provides us with focus on customer initial requirements to close the loop.

Other requirements of 9001 can be drawn into the IMS to varying degrees. I have focused on those that are critical.

Systems thinking is vital for successful innovation. If innovation is not integrated into the business system, it will die. With very little adaptation, the requirements of ISO 9001 provide an excellent framework for developing an innovation management system.

KEY POINTS

- A start-up business must have a management system if it is to succeed in the long term.
- In creating the management system, flexibility is essential.

Continued

Continued

- As organizations grow, they become less agile.
- A networked organization is more agile than a traditional organization.
- If an organization has grown too large and lost its agility, a spin-off solution may be needed in order to innovate.
- The spin-off should reduce hierarchy, encourage diversity, and be independent.
- The spin-off should develop its own IT and HR to encourage new knowledge and new behavior.
- The leader of the spin-off must reflect the new knowledge and new behavior.
- If the leader is recruited from outside, they may need to report to a higher level to bypass "roadblocks."
- A spin-off has separate business meetings from the parent company.
- The innovative organization measures nonfinancials.
- The brand of the new product or service must be meaningful to the customer.
- Use ISO 9001 in an agile form to create structure.

17

Measuring Innovation

If you don't keep score, it's only practice.

—Anonymous

I love sport and I love history. Both use measurement in different ways. Sport keeps score and measures through a simple numerical activity, and history measures by recording events. I see people struggle with measurement because they are often unsure of its purpose. We measure in order to learn.

People are often asked to record events, rather like a scribe in history, but are not included in the analysis of those events. It is the analysis where we convert data into knowledge:

- Data are just numbers.
- Information is patterns in the data.
- Knowledge is information that can be acted on.
- Innovation is produced from new knowledge.

The scribe often sees events that the pen does not capture. Of course, a good scribe takes notes, and I have already emphasized the importance of note taking in the world of the innovator. In the creative process, there are many unwritten events that must not be lost. They must be discussed and the learning acquired.

Measurement of the creative phase of innovation owes more to the measurement of history than to keeping score in sport. You measure in order to learn about your new opportunity and potential solutions. You also measure the creative process and how well it is performing, and what changes you need to make to it. Measurement of the execution phase of innovation is

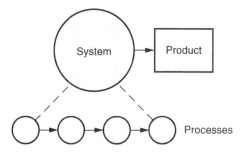

Figure 17.1 Levels of measurement.

more like traditional process measurement, or to be fair, "project measurement." In Chapters 6 through 10 you assessed your processes, and that was the first step toward a measurement plan.

There is a cruel twist for the innovator. One of your most important measures will be in order to "learn how well you learn."

The ultimate measure of whether your innovation is working is whether people buy your product or service. The classic metrics we hear referenced for innovation are "% of product less than three years old" or "new-product development life cycle." These are seriously lagging indicators and lack accountability. As a result, they lead to only small and non-radical innovation. We need leading indicators such as the extent to which our people are involved or the number of new ideas being generated. You measure where you need to improve, and you measure to drive results.

However, we must take a step back and assess at an organization or system level. In the same way as there are three levels of innovation, there are three levels of measurement: system, process, and product (see Figure 17.1).

We need to measure the system by looking at each of its elements. However, even within the system there is the critical first issue of *innovation aptitude*, or *areas of competence*, which we discussed in Chapter 3.

MEASURING INNOVATION APTITUDE

Innovation measurement starts by focusing on the aptitudes of the people who are required at each stage in the process. Successful innovation comes from the collective knowledge of the people in your organization. You must first measure "people strengths" so people can focus where they can make their best contribution.

Chapter 3 can help you assess not only where you will make your own best contribution, but also the contribution of the other people in your organization. It is imperative to maintain a mix of all types of people as you move through the innovation process. On the other hand, it would be fatal to be operating the development stage of the process without a majority of people who measured highest in that category.

The key roles in an innovative organization and in the innovation process match the stages in the process. Self-measurement will show you and your colleagues where you will make your best contribution, whether it's by:

- Generating opportunities (creators)

- Linking ideas to solutions (connectors)

- Turning ideas into practical solutions (developers)

- Delivering solutions and getting things done (doers)

MEASURING THE SYSTEM

We need to measure the system by looking at each of its elements and conducting an organizational assessment. You need to look at your organization and determine where it is deficient. There are six aspects of your organization you should evaluate. You need to ask questions like the following:

Leadership. Do you encourage risk taking, and do you have a commitment to innovation? Does your vision address innovation, and do your leaders and people search for new ideas?

Knowledge. Do you gather competitive intelligence, have knowledge sharing systems, and work with customers and suppliers? Do you have partners in key market areas?

Process. Do you have an innovation process that uses cross-functional teams?

People. Does a high proportion of employees have the capability to innovate, and do they engage in innovation? Is training in innovation offered? Do you encourage skunk works?

Culture. Is innovation recognized and rewarded, and do you document failures to learn? Do your hiring, training, and staff time allow for innovation?

Results. Do you win awards? What percentage of sales is from new products, and are products improved continuously? Does innovation achieve business goals?

Figure 17.2 is a simplified organizational assessment that will help you conduct this evaluation.

Leadership and Strategy	Strongly agree	Agree	Disagree	Strongly disagree
We make financial investment in innovation				
Our vision and mission includes innovation				
Our leaders take risks				
We target "white space" in the market				
We practice open market innovation				
Total (total each column)	4 ×	3 ×	2 ×	1 ×
Grand total =				

Key issues:

Knowledge management	Strongly agree	Agree	Disagree	Strongly disagree
We collaborate with customers and suppliers				
We share knowledge internally and externally				
We protect intellectual property				
We do not have high dependency on documented knowledge				
We learn from failure				
Total (total each column)	4 ×	3 ×	2 ×	1 ×
Grand total =				

Key issues:

Figure 17.2 Organizational assessment.

The assessment asks just five basic questions in each assessment area. This gives a maximum score of 20 points in each section. If you choose, you can multiply each section score by five to give a percentage score.

This assessment gives you a very high-level view of your organization, and areas where your score is low should be measured in far more detail. You have already assessed the next level of detail for *process* when you

Processes and system	Strongly agree	Agree	Disagree	Strongly disagree
We have a defined innovation system and process				
We generate many opportunities and solutions				
We have a selection process for new products				
We make our new products user-friendly				
We are fast to market				
Total (total each column)	4 ×	3 ×	2 ×	1 ×
Grand total =				

Key issues:

People involvement	Strongly agree	Agree	Disagree	Strongly disagree
People know their innovation aptitude				
We develop innovative competencies				
We have communities of innovation				
People are free to explore				
We have a balance of creators, connectors, developers, and doers				
Total (total each column)	4 ×	3 ×	2 ×	1 ×
Grand total =				

Key issues:

Figure 17.2 *Continued.*

Culture	Strongly agree	Agree	Disagree	Strongly disagree
We are not afraid to fail				
We recruit people who are innovative				
We manage a "loose/tight" culture				
We are an agile organization				
Innovative behavior is recognized and rewarded				
Total (total each column)	4 ×	3 ×	2 ×	1 ×
Grand total =				

Key issues:

Results	Strongly agree	Agree	Disagree	Strongly disagree
We measure learning				
We measure speed to market				
The majority of our products are less than five years old				
We do not need to have recalls or "service fixes"				
We win awards for innovation				
Total (total each column)	4 ×	3 ×	2 ×	1 ×
Grand total =				

Key issues:

Figure 17.2 *Continued.*

did your process assessments in the earlier chapters. However, you should at this stage list the key issues from each section of the assessment. This tells you where more measurement, and hence more work, is required (see Figure 17.3).

Your assessment is subjective. You have viewed the primary aspects of your organization and allocated an arbitrary score. However, if you have your full leadership team conduct the assessment, you will get an important consensus. We have carried out this consensus assessment with many

• Leadership and strategy
• Knowledge management
• Processes and system
• People involvement
• Culture
• Results

Figure 17.3 Key issues.

organizations, and it has helped them significantly in setting their innovation strategy. I can provide you with an online version if you contact me at pm@petermerrill.com.

MEASURING PROCESSES

Naturally, the essential rules of measurement apply. First, identify where your processes are weakest. You will find that from the assessment charts in Chapters 6–10. Good examples of measurement opportunities might be the ability of the initial opportunity stage to generate ideas or the ability of the final execution stage to get to market quickly. People usually go wrong by measuring the outcome or the product or the number of new products or patents produced. Other measurement examples might be the number of breakthrough ideas we develop, the performance of our relationships with inventors, partners, or suppliers as sources of original ideas, and the impact of our innovation efforts on our customers. These measures help us understand and improve the processes that support innovation.

As companies grow, they use measures as a way to evaluate performance. What starts as a few key measures leads to a "mountain of measurements." This mountain can lead to a focus on "completing the list" instead of using measurement to learn. Measures are fun to create and difficult to destroy. Review measures quarterly to see if they are still relevant. If they're not, eliminate them.

Earlier in this chapter you assessed where your company is on the innovation totem pole. Measure where you are weak. Drill down with your measures. You measure to learn about something you need to better understand and then to improve. Where your high-level measures don't tell you enough, go deeper. An initial measurement plan for your innovation process might look like Figure 17.4.

It is in the early stages of the innovation process where the measurement challenges arise. You need to focus on process inputs. The creative phase is about knowledge creation, and the metrics should focus first on

Process	Owner	Primary objective (example)	Performance measures (example)
Opportunity	Marketing	Find new opportunities	Percentage of new opportunities in new market
		Find market "white space"	
Solution	Design	"Out of box" solutions	Number of radical solutions
Selection	Leadership	Narrow focus	Number selected
		Take risk	Number of failures
Development	Development	Make user-friendly	Behavior change
		Rapid completion	Time to complete
Execution	Production	Speed of delivery	Speed
	Sales	Customer acceptance	Customer uptake

Figure 17.4 Innovation process measurement plan.

knowledge creation and identifying the customer experience. For example, at the opportunity stage you need to monitor market trends. Your market research people should do this. You need to measure both customer and market opportunities identified, and also the impact of those opportunities.

We measure the performance with partners, inventors, customers, and suppliers. As we go on to create solutions, we measure the number of new solutions created for each opportunity and get data on external and internal risk that attaches, and the radical extent of solutions.

The project tipping point is where the strategists reengage and address and mitigate risk. Radical innovations produce more profit but entail more risk. The risk assessment measures ROI against the radical nature of the idea and the anticipated speed to market. The new-product portfolio must balance short- and long-term time to market. We also conduct partner assessment and delivery chain risk assessment. We assess the partner's QMS as well as their deliverables.

A key strategic measure at this point is the stakeholder influence chart explained in Chapter 8 (Figure 8.5). People with vested interests can stop a project in its tracks. They need to be identified and their concerns addressed out in the open.

We are now at the execution phase of the project, and we now narrow focus. We move from a "loose" to "tight" mode of project operation and now move with speed. The development work comes first and is tasked to make the offering "user-friendly." Work by Daniel Kahneman for his 2002 Nobel Prize in economics shows that we need a tenfold increase in perceived value for our new idea to be accepted. We measure *ease of use* versus *radical nature*, and involve operations and sales for these measures.

The project team will usually break into task groups and involve the customer as we test solutions. Conventional project management metrics become critical here, measuring stage-gate time to completion and conducting frequent monitoring and review. Partnering and collaboration with suppliers and customers need to be monitored for effectiveness.

Risk will change as we gain knowledge, and the need for mitigation will change. We are constantly measuring test results and resource needs. Good resource management is vital to project success in this final phase. Design reviews need to be measuring rework, which is often disguised as "iteration."

MEASURING PRODUCT AND SERVICE

Measuring the later stages of your innovation process is not too difficult. You may well be doing this already. Software developers and IT technicians are on very familiar ground when measuring things like "first-time fix" and "errors per line of code." The same concepts apply at the development stage of any new product. At the execution stage, sales and production people are familiar with measuring speed to market, speed of acceptance, production errors, and service errors. My chapter on measurement in *Do It Right the Second Time* describes how to measure at these final stages of the innovation process.

THE MEASUREMENT STRATEGY

Remember, you are measuring process inputs to ensure that your innovation efforts will bear fruit. If your process inputs are not performing as you require, you need to act. Innovation measurement is about improving your ability to innovate.

Start by having your people measure their innovation strengths. Your people's knowledge is your primary tool as you initiate your innovation process. Identify the areas of your organization where innovation is weak. Remember, you measure to improve. Drill down in those weak areas and focus on measuring weak inputs to your innovation process. Measuring your outputs is useful, but only as a validation, after the fact, of whether your innovation process is working.

To be sure you have covered all the bases, review the model in Figure 17.5 and see if there are any areas of your management system where you are not monitoring or measuring. Monitoring means just "feeling the pulse." Measurement means collecting data and analyzing it to gain knowledge.

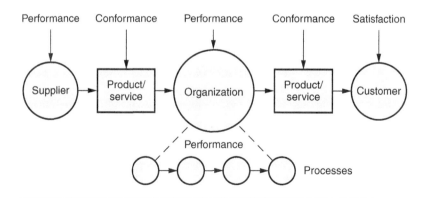

Figure 17.5 Areas for measurement.

As the innovative organization develops its ability to innovate, it must then expand its horizons to *open market innovation.*

> **KEY POINTS**
>
> - You measure innovation to see if it is working and to improve it.
> - The common mistake is to measure the product of the innovation process.
> - Measurement should focus on process and system.
> - The areas to evaluate in your system are leadership and strategy, knowledge management, processes and system, people involvement, culture, and results.
> - You should at some point conduct this assessment at a detailed level in your organization.
> - The organizational assessment enables you to focus on areas of weakness and build an action plan.
> - Process measurement should occur where the innovation process is weakest.
> - Focus on process inputs to avoid lagging indicators.
> - Develop a measurement plan.
> - Review the list of measurements on a regular basis and cull those that no longer have value.

18
Open Market Innovation

Knowledge is of two kinds. We know a subject ourselves, or we know where we can find information upon it.

—Samuel Johnson

My university thesis was on "The Plasma Jet as a Chemical Reactor." To this day I recall that my primary finding was that progress in the research and development of this concept had been painfully slow because of a lack of collaboration. The progress of the plasma jet, creating the fourth state of matter, never got off the ground because it was driven by a "trading" philosophy and not a research philosophy. In business we have an instinct to trade; in research our instinct is to share. I joined Courtaulds' research division after a school and university life of sharing knowledge. It seemed only natural to continue.

As a chemical engineer, I went into an industry where vertical integration and complete secrecy were the prevailing principles. Industrial espionage and patent protection were the norm. One of my own company's new product developments was an acrylic fiber. A sodium thiocyanate solution was the solvent for the acrylic, and it wasn't a great leap of science to move from the potassium-based salt of the competitor to the sodium thiocyanate that my own company used. R&D labs worked this way. R&D stood for *robbery* and *denial*. You could say it was innovation, but it certainly wasn't "breakthrough" innovation.

For breakthoughs, we must step out of the box. There is more knowledge "outside the box." There are limits to the innovation that can be achieved by working with today's customer. Once your innovation process is working, you need to extend it. Radical innovation comes easier with knowledge from outside. Products and services are developed faster and cheaper. If we want breakthroughs, the chances are that we will not have the necessary body of knowledge within the boundaries of our own group of people or even our own organization. We also know that the major innovation breakthroughs occur at the intersection of disciplines. However, people know that knowledge is power, so people retain knowledge and don't share it. On the other hand, in our world of astronomical knowledge growth it is no longer possible for one person to possess the knowledge that an organization needs.

One of the most convincing pieces of evidence for open market innovation came from IBM in their global CEO survey. The chart in Figure 18.1 shows how CEO respondents selected the most significant sources of innovative ideas.

The number alongside each category shows the percentage of respondents selecting that particular source of ideas as "primary." Each respondent was able to make up to three choices, and therefore a maximum theoretical score would be 300 percent. Not everyone made three choices, so you will see the sum total is 220%.

This survey shows overwhelming evidence that collaboration between employees, business partners, and customers produced the best innovation results. That is what open market innovation is all about.

The IBM global survey of CEOs also showed a devastating indictment of R&D labs and a remarkable shift to external sources of ideas and expertise. We need to recognize that the smartest people don't work for our own company, and with today's search abilities, finding smart people is much easier. The social interaction with others and the interaction with other disciplines are what generate both opportunities and solutions. You do not need to own the expertise.

This is what has led organizations such as Apple, Procter & Gamble, and IBM to develop their open market innovation. The premise of there being more knowledge outside the box is not new. In times of rapid innovation, knowledge lifetime shrinks. As far back as the Renaissance, which

Employees 41%	Consultants 22%	Internal sales 17%
Business partners 37%	Competitors 20%	R&D 16%
Customers 35%	Associations 18%	Academia 14%
113%	60%	47%

Figure 18.1 Sources of innovative ideas.

was a time of rapid innovation, the textile industry in Tuscany recognized that if you offer knowledge to others, you get new knowledge in return.

General Electric sourced a new design for the bracket that attaches its engines to aircraft through an online 3D printing challenge. The winner was an Indonesian engineer, whose blueprint for a stronger, lighter model was initially dismissed by GE's in-house team.[1]

HOW OPEN MARKET CAN WORK

Networks of innovation create far more potential solutions than the closed model of innovation.[2] Open market innovation has been especially successful in areas such as open source software development.

If you look at the stages in the innovation process, it becomes clearer who your external partners should be. Remember, breakthroughs come from the intersection of disciplines. Picking your partners is clearly a critical activity.

Stage 1: Creating the Opportunity

Sales people in your own organization mix with sales people in other industries. If they don't, they should. Your market research people should be looking at industry trends, but also mixing with customers and consumers. Remember, the mission here is to find an unsatisfied need. Remember also that the best opportunities come from secondary connections. This network needs to be dispersed.

Stage 2: Finding the Solution

This is where cross-connecting with other industries becomes essential. Remember, Henry Ford's idea for mass production came from a meat processing factory.

The link between processing lots of cars and processing lots of meat is in the word *processing*. The skill of exploration becomes more demanding here. The opportunity guys were explorers, but the solution guys have fewer choices.[3] Process thinking and conceptual thinking are vital.

Stage 3: Making It User-Friendly

This is where customer and consumer networking becomes critical. There are people downstream who will "tell it like it is." You need those people.

You are testing solutions to find what works. You don't have a lot of time. Knowing the right people to talk to means you must have been listening to what the stage 2 people were doing and prepared your network. These are your early-adopter customers that you are working with. You actually knew who they were back in stage 1 when the need was identified. This is also where you develop your supplier network

Stage 4: Execution

Here your network becomes about degrees of separation. One degree doesn't do it. The customer's customer, at the very least, and three degrees is preferred. What is holding up early adoption? One degree of separation will mask the truth.

One method of establishing partnerships is to use innovation go-betweens or headhunters to facilitate the exchange of sensitive information between companies. They connect interested companies with sources who can provide solutions to innovation challenges. You hire them to share information about your innovation needs with agents representing other companies and to help structure your engagement terms.

THE REWARDS

As one example, Procter & Gamble set a goal some years back for 50 percent of innovation to come from outside its organization. They were very open about the results. R&D productivity saw a 60% increase, their innovation success doubled, and R&D cost as a percentage of sales dropped from 4.8 percent to 3.4 percent. They produced 100 new products in two years, and their share price doubled.[4]

Apple took the iPod to market but PortalPlayer created the network or "iNet" that enabled the product to be developed and manufactured. They used vendors from different "ecosystems": disk drive from Singapore, semiconductor from Taiwan, and software from Bangalore.

There are many exciting stories of external collaboration. However, be careful. To quote one CEO, "Having a few beers together is not collaboration." Collaboration is a discipline. The IBM CEO survey showed that external collaborators outperformed the competition in both revenue growth and margin. Another interesting statistic in that survey was that public sector organizations collaborated externally to twice the extent of private sector organizations. James Moore in his book *The Death of Competition* talked of the importance of creating a business "ecosystem."

Since then, we have seen the concept of alliances evolve dramatically. However, we have at the same time seen that the majority of alliances are doomed to fail. How do we overcome that problem?

MAKING ALLIANCES WORK

A major reason for the failure of alliances is the focus on the legal and financial aspects of the alliance to the total exclusion of the relationship and resourcing aspects. Building trust is essential, and that trust builds when behavior follows expected patterns and patterns that the other party understands.

As a result, the method of working needs careful thought. Some people are comfortable with Skype, others prefer e-mail. Some are happy with conference calls, but others prefer regular face-to-face meetings.

It is far too easy to lay out generic principles in the early stages of relationship building. The principles must be specific. Avoid words like *review*, *appropriate*, *frequent*, or *necessary*. The decision-making process in each company needs to be understood by the respective partners. At what level are decisions made, and what are the steps in the process? What are the key decisions that are likely to be taken as the project moves forward?

In the same way that metrics for an innovative spin-off should be nonfinancial (see Chapter 16), there need to be nonfinancial leading indicators for the open market alliance. Obsession with short-term financials will kill the relationship.

Using the type of process measures described in Chapter 17, such as number of opportunities, number of solutions, and speed of bug elimination, will drive progress.

SEEK TO UNDERSTAND

An alliance is normally formed because each party brings an asset, whether it is market, intellectual property, or technology. Recognize that the other partner will not understand your technology if that is what you bring to the party. If the other partner brings market opportunity, you need to take time to understand that.

This is only the first step. Differences in culture are the big issue. What might at first be seen as "attention to detail" can rapidly become "slow and bureaucratic." A behavior that at first is seen as "fast response" can soon be seen as "reckless." In my own work in ISO, I work with cultures from across the planet. Europeans seem to want to debate forever, while North Americans race on with thoughtless abandon. East Asians are inscrutable, and people wonder what they are thinking. Taking time to form working relationships in our meetings is a critical part of our work.

Agree on how information will be shared and how problems are shared. Agree on approaches for both of these as well as for normal work activity. Problems may well arise from within your own company, and you have a responsibility to address them internally when that happens.

NETWORK ALLIANCES

After partnerships, the next level of behavior in open market innovation is *network innovation*. This is where your alliance is no longer with one or two other businesses, but can be with a myriad of other organizations. This is like forming a community of innovation, which was described in Chapter 12, but where the members of the community are all outside your organization. The issues of trust and relationship building now multiply exponentially.

You need a network coordinator who will recruit participants and talents just like in the community of innovation but on a larger scale. You need a "context for participation," such as chat rooms supported by Skype. There is a loose mode of operation in the early stages, and trust is key, but there are tight action points just as with internal innovation.

The coordinator should not be alone, but should have a small core team of three or four persons. The exact same issues are addressed as in forming alliances, but you probably need to keep agreements simple, as many participants in the network may well be "lone wolves." Clear timelines become important but should be balanced with freedom for participants between the "action points." Interaction between two or more members of the network is a clear challenge for the coordinator, and showing participants where they fit in the network is another part of building trust.

REAL-LIFE RESULTS

One of the more "fun" stories at P&G was of the "printed Pringle." We all know Pringles. They are potato chips for neat freaks. They certainly lack excitement. The innovative breakthrough was transferring the idea of the fortune cookie onto the Pringle. Fortune cookies would be boring like Pringles if they didn't have a message in each cookie. The old P&G would have spent five years developing the technology for printing vegetable dyes.

The new P&G bought it. They went on a global search and found a baker in Italy who had developed the technology for printing vegetable dyes onto bread. The baker loved the money and P&G got to market fast. They said at P&G, "We have replaced 'Not invented here' with 'Proudly found elsewhere.'"

IBM alphaWorks looks at open market innovation from a different perspective. Contrary to the image implied in the past by Apple advertisements, IBM is a highly innovative company.

You only have to look at their record going back to the invention of the ATM, their work on artificial intelligence, and the number of patents and scientific prizes the company has generated. So productive is their work in the field of software alone that they were feeling the frustration of their innovators in the 1990s as their ideas were failing to see the light of day.

The company set up the alphaWorks website, which offered free 90-day trials on unused software. The response was phenomenal as developers and businesses downloaded the software. Over 40 percent of the products going on the website are purchased, and IBM now earns $2 billion each year from royalties on its patents. But here is the twist: ideas came to IBM from their potential customers.

In the last three chapters, I introduced you to the implementation of innovation: the type of organization you need, how you measure progress, and the longer-term move to open market innovation. There is one other secret ingredient to implementation. Getting an early win and lighting the fire. I will explain this in the next chapter.

KEY POINTS

- Radical innovation is achieved more easily with ideas from outside your organization.
- Breakthroughs occur at the intersection of disciplines.
- Knowledge lifetime shrinks in periods of rapid innovation.
- External networks will change at different stages in the innovation process.
- Networks can be with customers, other industries, suppliers, or consumers.
- Collaboration must be disciplined.
- Alliances succeed when attention is given to relationships and resourcing, not just finances.

Continued

Continued

- Take time to understand how an alliance partner shares information and solves their problems.
- Network alliances are less formal.
- A network coordinator is a key player.
- The coordinator recruits participants, builds trust, creates the context for participation, and ensures tight action points.

19
Lighting the Fire

The time to hesitate is through
No time to wallow in the mire . . .
Come on baby, light my fire.

—Robby Krieger and Jim Morrison
The Doors, "Light My Fire"

I recall a company in England called Ashton Brothers. They made a product called Zorbit nappies, or what in North America would be called diapers. They had the most powerful brand in the baby care industry. They had been incredibly profitable, year in, year out, for nearly a century. The CEO was an accountant and rejected the innovative move to disposables. This could not be justified given their huge investment in terry weaving machines. That investment had to be recouped. Terry cloth was "what we do well" at Ashton Brothers.

My own company, part of the same corporation, looked on aghast. The global trend to disposables was moving like a juggernaut.

The Zorbit nappy business, after nearly a century of success, crashed and burned in two years. The brand disappeared overnight and became synonymous with "yesterday." The new generation of mothers no longer was interested in hanging a dozen bright white terry napkins on the washing line in the back yard. Washing nappies was no longer an "act of love" for the newborn. It was now domestic imprisonment.

You have to determine what social change is affecting your own market. You have to talk to your customer and use those questions from Chapter 6 to find out what is really bugging them. The evidence may eventually show in the declining revenue of one of your products, but by then it may well be too late. It was too late for Zorbit.

THE CHANGE TEAM

When Charles Darwin spoke to the problem of change 150 years ago, he said, "It is not the strongest of the species that will survive, nor the most intelligent, but those most able to respond to change." It is the ability to respond to change that is today's challenge.

The change team members are the change agents. The job of the team is twofold. First, to initiate the innovation process, but at the same time to drive the culture change that the organization must undergo. You will have a good sense of the changes needed from the organizational assessment in Chapter 17 on measurement.

You will recall from your own self-assessment earlier in the book that the change team also needs a balance of creators, connectors, developers, and doers.

The team should represent all functions from marketing through R&D to production and sales. The people on the team should represent all levels in the organization, but wherever the team member is not a functional head, the team member has to command the respect of their functional head. In an organization going through "innovate or die" or in survival mode, the team will need a high proportion of the senior management team. The team leader is either the company president or VP–marketing.

The biggest challenge the team has is to overcome resistance to change.

CHANGING TO AN INNOVATIVE CULTURE

Resistance to change to an innovative culture is especially common if the organization has a history of success. It will be necessary to create a sense of the need to change, and the strategy for handling change should follow the approach I outlined in Chapter 11.

The *road map for change* (see Figure 19.1) develops John Kotter's eight stages for leading change, using a pictorial approach. Change is not a linear process, but a picture enables us to capture a sense of sequence. You might, for example, create the change team before creating the sense of urgency. You may well raise the bar before creating critical mass. You will certainly be resourcing the process throughout, although this becomes critical when the mass is critical. You will notice I have split the model into *people* issues and *process* issues in line with the treatment in this book, but again, that is not absolute. Use this model as a simple visual to determine whether you are addressing all the key issues in the change process.

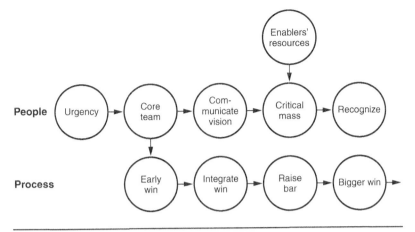

Figure 19.1 The road map to an innovative culture.

These are the stages in concept:

Create a sense of urgency. Identify a falling revenue item or market share with a key product or service or customer. Pick an item or a customer who is high-profile, and develop a sufficient desire for action. This is Kotter's "burning platform."

The change team. A change agent team must include people outside management and must be designed carefully. Their mission is to quickly create a critical mass of believers in the organization by replacing a dying product. The new product is the first visible evidence of developing an innovative culture.

Early win. The "low-hanging fruit" is captured. This is a product or service or customer opportunity with high benefit. Wins must be created, and not just based on hope. The win must produce a quick result and then be publicized.

Communicate vision. The change team must draw in the people by focusing on the two or three most important successes from the early win. These successes must be easily communicated in concise and simple wording. The team must keep repeating the successes. A "one-minute message" should be developed. Communicating the vision becomes a key task for the business leader.

Enable action. People must be given the authority and time to innovate. Leaders must find the obstacles and remove them. Compensation must be structured to support innovation.

Don't declare victory. The short-term win creates danger, as people relax. The outcome must be monitored and negative side effects dealt with. Significant culture change must be ensured before moving on to bigger issues.

Anchor new approach. Fundamental culture change is essential until innovation becomes business as usual. Culturally, the next generation must personify the innovation generation. Innovations must be explained and the participants recognized.

Recognize success. The behaviors described earlier are endorsed. Building trust reinforces new behaviors when they are based on the new values. Recognition and reward are aligned.

This gives you an outline of the change process; the business leader plays a key role in all of this. An excellent way of driving this change is by linking it with your project planning. Figure 19.2 shows a typical innovation project plan and change plan combined.

COMMUNICATE VISION

The change team must draw in the people by focusing on the two or three most important successes. These successes must be easily communicated in concise and simple wording. They must keep repeating the successes.

A valuable tool in doing this is the "one-minute message." Phil Crosby used to call this the "elevator speech." The term comes from when Phil was VP of quality at ITT. One day he entered the elevator at the ITT building along with Harold Geneen, the CEO of ITT. Harold asked Phil the simple question, "How's quality, Phil?" In the 60 seconds it took the elevator to reach the top floor of the ITT building, Phil Crosby told Geneen what was wrong with quality at ITT, how to solve it, how much money could be made, and importantly, what he wanted Geneen to do. The "motivational sequence" that I mentioned earlier flows like this:

1. Gain attention
2. Visualize need
3. Answer need
4. Visualize result
5. Call to action

"Change" team plan	Jan	Feb	Mar	Apr	May	Jun	Jul	Aug	Sep	Oct	Nov	Dec	Jan	Feb	Mar	Apr	May	Jun
Planning																		
Organizational assessment																		
Management workshop																		
Change team																		
Opportunities																		
Opportunity selection																		
Project plan																		
Initiation																		
Short-term win																		
Process development																		
Develop vision																		
Creativity workshops																		
Communities of practice																		
Customer focus																		
Build critical mass																		
Implementation																		
Strategic planning and selection																		
Product development																		
Value proposition																		
Project reviews																		
Launch																		
New product deployment																		

Figure 19.2 The change team plan.

The elevator speech in your own company will be entirely different from the speech in other companies. If your mission is to recruit people to join communities of innovation, it could look like this:

> We have been making Z for nearly a hundred years, but we all know from the global data that the product will die in the next two years. We need to replace this product with one that meets the needs of tomorrow's family and not today's. We have identified three critical projects that need immediate execution to achieve that change. If we execute these projects successfully, we will open new business with 10 major new customers over the next year. I would like you to join the community of innovation.

Less than 100 words is your target. Every member of the change team must have the team's elevator speech in their pocket.

Building innovation into the company's vision is a task led by the president.

BUILD CRITICAL MASS

Communicating vision is one of the toughest tasks. Seventy-five percent of the people must become believers in the first year, a 3:1 ratio. Storytelling, which I described in Chapter 13, is one of the key skills here. Stephen Covey, in his book *The 7 Habits of Highly Effective People*, describes the second habit, "Begin with the end in mind." He describes the building and sharing of a vision.

The strengthening of the stages in the innovation process is the job of the senior person or persons in that functional area: at the *creating* stage the VP–marketing, *connecting* stage the VP–research or quality, *developing* stage the VP–R&D, and the *doing* stage will have both operations and sales working in tandem. They are key players in building critical mass.

After the early win, the lessons learned need to be applied to future projects. The organizational assessment should be revisited with a fresh perspective. Now is also the time to clearly define your innovation process. This is in anticipation of raising the bar and ensuring that in the future people follow the process.

One of the issues I talked about earlier in this book is how the senior people acquire mind-sets over the years, and even though they accept intellectually the need to eliminate old products, you're probably talking about something that was once their "baby." Passive aggression, enemies "lying in the weeds," and silent resistance are the greatest enemies of change. These must be addressed and brought out in the open. This is the antithesis of vision. It is finding out what people want to hang on to and addressing the issue. I also addressed this resistance to change with the influence chart in Chapter 8.

ENABLE ACTION

People must be given the authority and time to innovate. Leaders must find the obstacles and remove them. Compensation should be structured to support innovation. Remember, the secret is to reduce hierarchy, increase autonomy, and encourage diversity.

The popular view of innovation is that it is just "technical stuff." It is actually far more "people stuff." The leadership role now starts to entail not just *commitment* but also *involvement*. The senior leadership will need to deal with politics and kill any attempts to stop change. They will need to ensure that skunk works activities are properly resourced. Resourcing means ensuring that people truly have time to do the project work. At this stage, we are starting to see ownership of innovation move away from the change team and into the hands of senior leaders as we build critical mass.

RAISE THE BAR

The short-term win creates danger, as people relax. There are more ideas outside the box. Open market innovation is your next destination. When you "raise the bar," you enter into more ambitious projects as well as engage in open market innovation with your business partners.

Continual culture change is essential until innovation becomes business as usual. Culturally, the "next generation" must personify innovation. New people entering the organization must possess the values and behaviors of innovation described in Chapter 11. The innovations you produce must be explained to both these people and people already in the organization.

INTEGRATE THE WIN AND THE CHANGE

Linkage of the innovation process to the longer-term business strategy becomes vital at this stage. The project *selection* stage of the process is owned by the leadership and strategic planners. The business now needs to look outside for strategic partners as the bar is raised.

For information to flow, we need processes, people, and technology to come together as a management system. But this system must be flexible. Innovation is seriously inhibited by a lack of organizational agility.

The organizational assessment is repeated in full, and the defined innovation processes are also reassessed for weaknesses.

RECOGNIZE SUCCESS

Throughout the stages I have described, a comprehensive recognition system needs to have been operating. I described this in Chapter 11. The behaviors to be endorsed are identified, and new behaviors are recognized, not heroic defenses of the old. Building trust reinforces new behaviors when they are based on the new values. Recognition and reward must be aligned.

An early win is critical if you are going to get started and are going to make people believe in your desire to innovate. Find that mature or declining offering and use it to ensure the 'win' and light the burning platform.

KEY POINTS

- Using change agents is a proven way of initiating the innovation culture change.
- A representative group of change agents needs to be carefully chosen.
- Create a road map and a project plan for change.
- Focus on a product, service, or market segment with declining revenue to gain attention for the need to change.
- An early win is essential.
- Create a "one-minute message" to communicate the need for change.
- The business leader is responsible to communicate the vision of innovation.
- The business leader must ensure that people are given time and that roadblocks are removed.
- Innovation must be integrated into the business strategy.
- The business leader must be high-profile in recognizing innovative behavior.

20
Creating the Future

The best way to predict the future is to create it.

—Alan Kay

Innovation has moved up everyone's agenda dramatically in the last decade. This is because we are all becoming increasingly concerned about the future. This is why we are the "innovation generation." We all see the globalization that has been gathering speed over the last 50 years, which shows itself in high-speed communications, an increasing and aging population, and, sadly, the deterioration of the natural habitat we call planet Earth.

Happily, there has so far been no limit to human ingenuity in addressing these issues.

Innovation needs people with passion. I invite you to "step out of the box" and release your natural passion. The previous chapter was very much about process; this last chapter provides that ever-important balance with people.

Alan Kay's saying is so true—"The best way to predict the future is to create it." This book has been about how you and I will create that future.

Innovation is about answering the needs of tomorrow. It is not about technology, it is about the needs of people like you and me, and technology is only one way of answering those needs. It is need, not technology, that drives innovation.

There are other myths in the world that belie the truth of innovation. It is the "magic touch" between you and others that leads to radical innovation. We understand far better today the importance of sharing collective knowledge. We understand far better the importance of diversity and the importance of giving ourselves time and space to share knowledge.

The common myth is that the greatest challenge for innovation is that of creativity. In truth, just as big a challenge is getting the new idea "out there." That is because when it is an idea, it is not yet a product or service. The organizations we have created all too often get in the way of execution. Lack of organizational agility is the greatest barrier to innovation.

SO, WHAT DO WE DO?

The primary symptom of a dysfunctional organization is the failure to meet the customer's needs and expectations—the failure to deliver quality. The last two decades of the twentieth century saw people understanding far better how groups of people—we call them organizations—work well together. This led to the emergence of quality management as a way to make organizations more effective at delivering quality.

Our problem is that many quality managers have become focused on correcting the problems of yesterday's failed delivery. They have taken their eye off the ball, and that ball is the customer needs of tomorrow.

This is like a racing driver who keeps tinkering with the engine of the car but never actually takes part in the race.

In the face of global trends, it is tempting to think there is little we can do to address the consequences of our mushrooming and aging population, global warming and pollution, and the inequality of the world's wealth.

The temptation is to be like an ostrich, ignore the realities of the world, and keep tinkering with the quality management system. It's time to stop tinkering and take the vehicle out on the road. You have seen that an effective QMS is fundamental to successful innovation, so let's move on and join the race and innovate.

FULFILLMENT FOR THE INNOVATOR

Far too much of business has become obsessed with the score instead of the game. If you play golf or any other sport, you know that obsessing about the score will destroy your game, whereas focusing on the game will bring about the result you want.

The result we want is a strong and healthy organization with a long-term future, and we know that the game is fulfilling the needs of both today's and, importantly, tomorrow's customer. Playing a good game is one of the most personally satisfying things anyone can do, and that game can be any activity that we engage in. Innovation, at whatever stage you participate, is one of the most satisfying and fulfilling of all "games."

WHERE DO I BEGIN?

You start with a personal assessment, and then follow it with an assessment at an organizational level.

For successful innovation, you must know the aptitudes of the people in your organization. You already know whether you are a creator or connector, a developer or a doer. But what of your colleagues, where are their aptitudes? At a personal level, you should also start to develop and operate your network. Find out the aptitudes of other people in your network by asking them to complete the self-assessment, and explain to them what it means.

Engage your people where their best contributions can be made, and develop the "new behaviors" I have described. These new behaviors release knowledge. You need to release the *wisdom of crowds* through *communities of innovation* in which a key feature is dialogue. This engages the tacit knowledge in your colleagues, which in turn develops your ability to innovate. You can then develop more communities of innovation to build your critical mass as you move toward an innovative culture.

Looking at your organization, an in-depth assessment, going deeper than the assessment in this book, will point you to where your issues lie and the actions you need to take. You need to look at your own organization and decide where it needs to change and how to make those changes happen. You now have an understanding of the innovation process and the different roles in the process. You should address your innovation process in detail using the assessment tool in Chapters 6 through 10. Do this in discussion with your colleagues, not in isolation. This will point you to some immediate actions you can take with your innovation process. You may well find that you need to look afresh at how marketing, research, development, operations, and sales are structured in your organization.

Moving to a higher level, the innovation process of the organization needs to be integrated with the strategy of the organization. Strategy is about finding "green field" market space and about engaging in open market innovation. The third aspect of strategy must be the development of an innovative culture and those critical skills and behaviors in the early stages of the innovation process.

Culture and behavior have been addressed in major part in this book. At a strategic level, the critical issue is endorsing the behaviors you wish to encourage in recruitment, personnel assessments, and through recognition.

Looking to the future, one issue that is often overlooked with culture change is the "rite of passage." I have not dwelt on this, but it should not be overlooked. Letting go of the old and embracing the new is something that cultures through the ages have recognized as essential, whether it is the coronation or inauguration of a new head of state or the coming of age of a member of our own family. This is not just a celebration of the new, this is also a release of the old ways, and this must happen as you move to an innovative culture. Time scale? Think 18 months down the road for your "innovation celebration."

THE RESULT

Self-fulfillment, whether at a personal or an organizational level, is the best thing that can happen in life, and yet it eludes the majority of people and the majority of organizations. Ask Maslow! Innovation enables that fulfillment better than any activity you will engage in.

We all have a creative need, and yet the needs of the modern organization have all too frequently subverted this personal need. Business is not a battle; it should be about collaboration, not conflict, and be about growth, not greed.

Follow the path I have described, and you will become both an individual and an organization in which the left and the right brain are balanced, where there is trust and honesty, and you will achieve fulfillment.

WHAT DO I DO NOW?

Know yourself and know your colleagues. Know your organization, use the self-assessment. Do the process and organizational assessment.

Educate and explore. Host a workshop so that you can talk with others in your company who are interested in innovating. Select those new places to go and new people to meet. Take notes.

Get an early win, for yourself and your organization. I already talked about how to do this for your organization. But what about yourself? Realizing your personal potential is essential. Use your own creativity in order to do that. Your early win will involve you doing something you have never done before, and maybe were frightened to try.

Explore, observe, take notes, share ideas, and then try that something new. Don't be afraid to fail. Whether you succeed or fail, congratulate yourself for doing something you have not done before. Give a gift to yourself. The gift to yourself should be something that you will remember whether it is an experience or something you can use. The fulfillment when you try something new and succeed will be an excitement that spurs you to try other new things.

I invite you to step out of the box and realize your own potential as an innovator. Join the "innovation generation" but once you have joined, you must remember "innovation never stops."

Endnotes

Chapter 1

1. Michael Porter, *Creating and Sustaining Performance by Competitive Advantage* (New York: The Free Press, 1980).
2. Rita Gunther McGrath, *The End of Competitive Advantage: How to Keep Your Strategy Moving as Fast as Your Business* (Boston: Harvard Business Review Press, 2013).
3. Bart Stuck and Michael Weingarten, *Innovation And Profitability*, (Boston: Signal Lake, 2004).
4. Jane C. Linder, "Measuring Profitable Growth and Innovation," *Accenture Survey* (January 2006).

Chapter 2

1. Frank Zollner, *Leonardo da Vinci* (Cologne: Taschen, 2006).
2. James Gleick, *Chaos: Making a New Science* (New York: Penguin, 2008).
3. Herve Mignot, leader of the French delegation to ISO/TC 176, in conversation with the author.
4. Albert-László Barabási, *Linked* (London: Penguin, 2002).
5. Ikujiro Nonaka and Hirotaka Takeuchi, *The Knowledge Creating Company* (New York: Oxford University Press, 1995).
6. Ross Dawson, *Developing Knowledge-Based Client Relationships* (Oxford: Butterworth, 2000).
7. James Surowiecki, *The Wisdom of Crowds* (New York: Doubleday, 2004).

Chapter 4

1. Peter Senge, *The Fifth Discipline* (New York: Doubleday, 1990).
2. IBM, *Expanding the Innovation Horizon: 2006 Global CEO Study* (Somers, NY: IBM, 2006).

Chapter 5

1. John Naisbitt, *Megatrends: Ten New Directions Transforming Our Lives* (New York: Warner Books, 1984).
2. Donnalla H. Meadows, Dennis L. Meadows, Jørgen Randers, and William W. Behrens III, *The Limits to Growth: A Report for the Club of Rome's Project on the Predicament of Mankind* (Washington, D.C.: Potomac Associates, 1974).
3. W. Chan Kim and Renee Mauborgne, *Blue Ocean Strategy* (Boston: Harvard Business Review Press, 2005).
4. Michael Porter, *Creating and Sustaining Performance by Competitive Advantage* (New York: The Free Press, 1980).
5. Ibid.
6. Jim Collins, *Good to Great* (New York: HarperCollins, 2001).
7. Rita Gunther McGrath, *The End of Competitive Advantage: How to Keep Your Strategy Moving as Fast as Your Business* (Boston: Harvard Business Review Press, 2013).

Chapter 6

1. Michael Webber, "How to Make Our Food System More Energy Efficient," *Scientific American* (January 2012).
2. Ibid.
3. George Day and Paul J. H. Schoemaker, "Scanning the Periphery," *Harvard Business Review* (November, 2005).
4. "Vodafone and TomTom Team on Cellular Nav System," PC Pro, October 27, 2006, accessed April 23, 2015, http://www.pcpro.co.uk/news/hardware/96664/vodafone-and-tomtom-team-on-cellular-nav-system.
5. Rashik Parmar, Ian Mackenzie, David Cohn, and David Gann, "The New Patterns of Innovation," *Harvard Business Review* (January/February 2014).

Chapter 7

1. Scott Page, *The Difference: How the Power of Diversity Creates Better Groups, Firms, Schools, and Societies* (Princeton, NJ: Princeton University Press, 2007).
2. Jonah Lehrer, "Groupthink," *The New Yorker* (January 30, 2012).
3. Keith Sawyer, *Group Genius* (New York: Basic Books, 2007).
4. Jacob Goldenberg and Roni Horowitz, "Finding Your Innovation Sweet Spot," *Harvard Business Review* (March 2003).

Chapter 8

1. Malcolm Gladwell, *The Tipping Point* (New York: Little, Brown & Co., 2000).

2. Aaron De Smet, Mark Loch, and Bill Schaninger, *Anatomy of a Healthy Corporation* (New York: McKinsey, 2007).
3. Bansi Nagji and Geoff Tuff, "Managing Your Innovation Portfolio," *Harvard Business Review* (May 2012).
4. George Day, "Is It Real?" *Harvard Business Review* (December 2007).

Chapter 9

1. Timberland 3D Printing. The Timberland Company, Stratham, NH. http://www.scansourceposbarcode.com/~/media/pos-and-barcode-us/Programs/scansource-3d/training/documents/improving_design_cycle_102314.pdf.
2. William Bridges, *Managing Transitions* (Cambridge, MA: Da Capo, 2003).
3. Daniel Kahneman and Amos Tversky, *Choices, Values, and Frames* (Cambridge: Cambridge University Press, 2000).

Chapter 10

1. James C. Anderson, "Customer Value Propositions in Business Markets," *Harvard Business Review* (March 2006).
2. "GetThere and IBM," getthere.com, accessed April 24, 2015, http://www.getthere.com/resources/documents/GT_IBM.pdf.

Chapter 11

1. Terrence Deal and Allan Kennedy, *Corporate Cultures* (New York: Perseus, 1982).
2. Ibid.
3. Keith Ferrazzi, *Never Eat Alone* (New York: Doubleday, 2005).
4. Peter Merrill, *Do It Right the Second Time* (Portland: Productivity Press, 1997).
5. John Kotter, *Leading Change* (Boston: Harvard Business Review Press, 1996).

Chapter 12

1. Lynda Gratton and Tamara Erickson, "Eight Ways to Build Collaborative Teams," *Harvard Business Review* (2007).
2. "Linus Torvalds," The History of Computing Project, last modified November 19, 2007, accessed April 23, 2015, http://www.thocp.net/biographies/torvalds_linus.html.

Chapter 13

1. Stephen R. Covey, *The 7 Habits of Highly Effective People* (New York: Simon and Schuster, 1989).

Chapter 14

1. Brian Uzzi and Shannon Dunlap, "How to Build your Network," *Harvard Business Review* (December 2005).

Chapter 15

1. Jim Collins, *Good to Great* (New York: HarperCollins, 2001).
2. International Organization for Standardization, ISO 10006:2003, *Quality management in projects* (Geneva: ISO, 2003).

Chapter 16

1. Vijay Govindarajan and Chris Trimble, "Building Breakthrough Businesses within Established Organizations," *Harvard Business Review* (May 2005).

Chapter 18

1. "Jet Engine Bracket from Indonesia Wins 3D Printing Challenge," GE.com, December 11, 2013, accessed April 23, 2015, http://www.gereports.com/post/77131235083/jet-engine-bracket-from-indonesia-wins-3d-printing.
2. Tuba Ustener and David Godes, "Better Sales Networks," *Harvard Business Review* (July 2006).
3. James Moore, *The Death of Competition* (New York: HarperCollins, 1996).
4. Larry Huston and Nabil Sakkab, "Connect and Develop—Procter & Gamble Model for Innovation," *Harvard Business Review* (March 2006).

Index

A

action
 enabling, 221
 taking, 175
additive manufacturing, 102
Agile Manifesto, 44
Agile Urban Logistics project, UK, 63
Alexander the Great, 23
alliances
 network, 212
 successful, 211
Alltree, Andy, 133
alphaWorks, IBM, 38, 213
Altshuller, Genrich, 77
Amazon.com, 4
Apple Inc., 7, 88, 208, 210
Archimedes, 25–26, 72, 126
Aristotle, 23
Ashton Brothers, 215
ASQ Innovation Interest Group, 128, 135

B

Barabási, Albert-László, 20, 157
Barker, Joel Arthur, 3
behavior
 changing, 130–31
 and culture, xxii, 123–32
 and knowledge, 185
 recognition of, 130
 and values, 125–30
behavior change, in user, 104
Bell, Alexander Graham, xix, 91, 92, 97
benefit
 versus features, 193
 versus value, 12
Biddle, Charlie, 127
Big Data, 63
BlackBerry, 3
blue ocean, market, 49
Box, George, 101
brainstorming, 74
brand identity, 185–86
Branson, Richard, 79
British Airways, 78–79
British Mini, 92
business, primary process areas, 6
butterfly effect, 17

C

Canadian Standards Association (CSA), 86–87
Capital One Corporation, 86
challenge, versus vision, 172

change
    resistance to, 105–6, 115
    using project to initiate, 175–77
change team, 134, 216
Chaos: Making a New Science, 17
chemical industry, 48, 207
China, 5, 7
Christy, Henry, 109–10
Christy, William Miller, xvii, 109
Christy Home Textiles, xvii, 98, 109–10, 111, 186
Cirque du Soleil, 7, 41, 53, 60, 62, 73
closed network, 160
cluster network, 160
collaboration
    in community of innovation, 138–39
    as value for innovation, 128
collective knowledge, 24–25
    leveraging, 16–17
Columbo, 147
community of innovation, 128, 134–35, 225
competence
    definition, 18
    innovation competencies, 146
    innovator, 143–54
connecting, in innovation process, 71–72
connector, role in innovation, 31, 32–33
content management, 24
continual improvement, 192
convention, and innovation, 8–9
Corning Incorporated, 184
Courtaulds, xvii, 7, 69, 155–56
creative problem solving, 149
creativity
    and innovation, 11–13, 33–34
    leading, 168–69
creator, role in innovation, 31, 32
Crest toothpaste, 186
critical mass, building, 220–21
Crosby, Philip, xvii, 9, 39, 130, 218
crowdsourcing, 165
culture
    and behavior, 123–32
    developing, 167–68
    of innovation, xxii, 28
        changing to, 216–18

organizational, 28–29, 121–77
understanding differences in, 211–12
customer needs, 53–55, 61, 63–64
customer pain statement, 49, 64
customer value proposition, 112–13, 193
    developing, 114–15

D

Da Vinci, Leonardo, xix, 15–16, 148
Da Vinci Code, The (movie), 15
Darwin, Charles, xvi, 5, 216
data, for innovation, 80–81
Data General, 139
Dawes, William, 158
de Bono, Edward, 75
de Guess, Arie, 15
decision making, leader's role in, 174
delivery, stage of innovation process, 42, 109–19
    team change, 141
developer, role in innovation 31, 34
development
    and queuing theory, 99–100
    stage of innovation process, 42
    steps in, 100–102
Digital Equipment Corporation, 139
disruptive innovation, 4–5, 13
doer, role in innovation, 31, 35
dragon boating, 181
Dvorak keyboard, 105

E

early win, 137–38
Eastman, George, 85
Edison, Thomas, xix, 10, 42, 74, 91, 97, 173
Einstein, Albert, 69, 73, 152, 192
Eisenhower, Dwight D., xxiii, 146
e-learning, 4
elevator speech, 218–20
embracing failure
    as innovation competency, 150
    as value for innovation, 129–30

Index   235

entrepreneurship, and innovation, 173–74
environmental scan, 49, 62–63
epiphany, myth of, 72–74
Euler, Leonhard, 157
Euler's network theory, 17, 157
experience, 143
experimentation, as value for innovation, 128–29
explicit knowledge, 21, 22, 23, 24
exploration
    as innovation competency, 146
    as value for innovation, 125–26

F

Facebook, 65, 157
failure, embracing
    as innovation competency, 150
    as value for innovation, 129–30
features, versus benefits, 193
Ferguson, Niall, 146
five forces model, Porter's, 7, 50–52
five generic innovation patterns, 77–78
Fleming, Alexander, 84
Ford, Henry, xix, 9, 10, 40, 41, 64, 73, 74, 125, 209
friend networks, 156–57
Frost, Robert, 83
Fry, Art, 38
fulfillment
    of innovator, 224–25
    personal, 226
future, creating, 4–5, 223–27
futurism, 48–49

G

GE (General Electric), 53, 209
Geneen, Harold, 218
Gleick's chaos theory, 17
Goldsmith, James, 47
Good Hope Hospital, 16, 146
Google, xv, 99, 105–6, 126
Gore, Al, 86
Gray, Elisha, 97
Great Exhibition of 1851, 110

green field, market space, 49, 61
Groupon, 50, 53, 61, 71

H

Hakes, Chris, 128
hedgehog principle, Collins's, 51
higher education, innovation in, 4
Hockney, David, 158
Hubbard, Gardiner, 97

I

IBM
    alphaWorks, 38, 213
    Global Expense Reporting Solutions, 117–19
IBM Global CEO Study, 45, 97, 168, 208, 210
ideation 74, 75–76, 149
IKEA, 40, 73
implementation, of innovation process, 179–227
Inconvenient Truth, An, 87
India, 5, 7
individual learning, 18
influence chart, 93–94, 203
information flow, 18–20
infrastructure, for product offering, 92
innovation
    competencies, 146
    convention and, 8–9
    creativity and, 11–13, 33–34
    culture, xxii
    disruptive, 4–5, 13
    essentials of, 1–56
    fundamentals of, 3–14
    history of xix–xx
    versus innovation management, xv
    leading, 165–77
    levels of, 45
    measuring, 195–205
    myths of, 9
    quality and, 5–7
    quality management system and, 43–44
    starting point for, 191–92
    structuring for, 187

innovation aptitude, measuring, 196–97
innovation leader, role of, 165–66
innovation management, versus innovation, xv
innovation management system (IMS), 17, 45
innovation process, xxi–xxii, 9–11, 57–119
  drivers of, 39
  initiating, 225–26, 226–27
  leading, 169–71
  measuring, 201–3
  networks in, 158–59
  roles in, xx, 27–36
  two phases of, 81
innovation projects, 135–36
innovation review, 175
innovation strategy, 47–56
innovation teamwork, 133–42
innovative culture, changing to, 216–18
innovative organization, 181–94
innovative thinking (test), 150
innovator
  competent, 143–54
  fulfillment for, 224–25
  mind of, 8
interaction
  as innovation competency, 146
  as value for innovation, 127
Invitation Digital Ltd., 50
iPhone, 92
iPod, 106, 210
ISO 9001:2015, 182, 187–88, 188–93
ISO 10006:2003 standard, 176
ISO 50500 standard, xv, 45, 187
ISO/TC 279 technical committee, 45, 187
Issigonis, Alec, 92
ITT Corporation, 218
iTunes, 92

J

Jobs, Steve, 85, 173, 185
Johnson, Samuel, 207

K

Kahneman, Daniel, 85, 105, 203
Kavanaugh, Gerry, 181–82
Kennedy Space Center, 151
key process indicators (KPIs), 186–87
Kipling, Rudyard, 143, 147
knowledge
  and behavior, 185
  collective, 24–25
    leveraging, 16–17
  growing, 22–24
  need for, 15–26
  as resource, 172
knowledge management (KM), xx, 21–22
  cycle, 22–24
Kodak (Eastman Kodak) Company, 4, 60, 85, 106
Kotter, John, 216
Krieger, Robby, 215

L

Laforet, Jim, 79
Laliberté, Guy, 60, 62, 71
Lao Tzu, 165
Larsen, Norm, 129
Last Supper, The, 15
lateral thinking, 75
Lauren, Ralph, 49
leader, role in innovation process, 165–66, 169–71
leadership, under ISO 9001:2015, 189
leadership workshop, 143–44
learning, individual and organizational, 18
Lewis, C. S., 158
Li'l Abner, 134
LinkedIn, 157
Linux software, 139
listening, as innovation competency, 147
loose mode of innovation, 39, 60–61
  switching from, 102–4, 139

M

Machiavelli, Niccolò, 123
macho culture, 28, 120
Madonna and Child, 15–16
management workshop, 144
Mann Deshi Mahila Sahakari Bank (MDMSB), 93
market space, uncontested, 49, 61
marketing, 185–86
measurement
   of innovation, 195–205
   of innovation processes, 201–3
   levels of, 196
   of product and service, 203–4
   strategy, 204–5
   of system, 197–201
mega trends, 48–49
Michelangelo, xix
Michelin run flat tire, 91–92, 114–15
Microsoft Corporation, 104
Microsoft Office, 104
Milliken & Company, 7, 53
mind map, 117
   software, 148
Mindjet software, 148
minor innovation, 88–91
   versus radical innovation, 88
mirroring, 159–61
Mission: Impossible III, 102
Morgan Stanley, 13
Morrison, Jim, 215
Mozart, Wolfgang Amadeus, 157–58

N

network alliances, 212
network innovation, 212
network theory, 20–21
networking, 25, 149, 155–63
networks
   and innovation process, 158–59
   structure, 21
new product introduction, 115–17
Nokia, 4
Nooyi, Indra, 189
note taking
   as innovation competency, 147–49
   as value for innovation, 127–28

O

observation, as value for innovation, 127–28
one-minute message, 218–20
online universities, 4
open market innovation, 207–14
   real-life results, 212–13
   rewards of, 210–11
   stages of, 209–10
   understanding in, 211–12
open network, 160
Open University, 4
opportunity
   identifying, 40–41, 49–50, 59–67
   under ISO 9001:2015, 189–90
organization
   flexible, 182
   growing, 182–83
   innovative, 181–94
   layers of, 182–83
   measurement of, 197–201
   spin-off, 184
organizational assessment, 197–201, 225
organizational learning, 18
organizational structure, and information flow, 18

P

Pauling, Linus, xxi, 37, 70, 75, 136, 158
PepsiCo, 189
Pepys, Samuel, 158
performance review, 131
personal knowledge network, 162–63
photography, innovation in, 60
Plato, 59
Plautus, 133
Polaroid Corporation, 60
Polo brand, 49–50
Portal Player, 210

Porter, Michael, 7, 50
Post-It notes, 38, 128–29, 148
Pringle's potato chips, 212
problem definition, 70–71
problem solving, creative, 149
process culture, 124
process of innovation, xxi–xxii, 9–11
process map, 18
process mapping, 18–20
Procter & Gamble, 186, 208, 210, 212–13
product, measurement of, 203–4
project, using to initiate change, 175–77
project team, 134, 135–41
   leading, 136–37

Q

quality
   and innovation, 5–7
   versus quality management, xv
quality management, 6–7
   versus quality, xv
quality management system (QMS)
   and innovation, 43–44
   managing, 192–93
Queen Victoria, 110
questioning, as innovation competency, 147

R

radical innovation, versus minor innovation, 88
"raising the bar," 221
R&D (research and development), xv, 97, 99, 210
recruitment, employee, 131–32
relaxation, for releasing knowledge, 25–26
Research In Motion, 3
Reshef, Shai, 4
resources, providing, 172–73
return on investment (ROI), 85, 86, 89, 103, 174
Revere, Paul, 158

review, in innovation process, 175
risk
   evaluating, 190–91
   internal and external, 91–92
   under ISO 9001:2015, 189–90
   and minor innovations, 89
   and social responsibility, 93
risk mitigation, 92–93
risk taking, 83–84
"robbery and denial" R&D, 98, 207
roles, in innovation process, 27–36
   assessing, 29–31

S

sacred product attributes, removing, 73, 78–79
Salk vaccine, 93
service, measurement of, 203–4
short-term thinking, 86–88
Silicon Valley, 139
Silver, Spencer, 38
skills, for innovation, 53
Škoda auto company, 87
skunk works, 134
Skype, 211
SMART (SCHMART) goals, 141
Smith, Trevor, 106
social network analysis, 21
social networking, 65
social responsibility, 48
   risk and, 93
sock exercise, 74–76
solutions
   connecting to, 69–82
   developing, 97–107
   finding, 192
   selecting, 33–34, 41, 84–85
Sony Corporation, 60
Southwest Airlines, 40–41, 73, 79
Spectra Energy Corp., 79
spin-off organization, 184
Stalin, Joseph, 77
Starbucks Coffee, 139
statement of value, 113
storytelling, as innovation competency, 152–53
strategy, 7–8, 225

Index 239

setting, 188–89
subconscious knowledge, 11
subprime mortgage problem, 93
success
   and networking, 157–58
   recognizing, 222
system, measurement of, 197–201
systematic inventive thinking, 77

T

tacit knowledge, 21, 22, 23, 24
team culture, 28
   shift in, 139–40
teamwork, innovation, 133–42
teams, types of, 134
technology, as innovation
      competency, 147–49
thinking space, as innovation
      competency, 151
Threadless, 65
three-dimensional risk diagram, 103
3M Company, 38, 99
Tide detergent, 186
tight mode of innovation, 39
   change to, 102–4, 139, 140–41
Timberland Company, 102
time, gift of, as innovation
      competency, 151–52
tipping point, in innovation process,
      41–42, 83–95
Tipping Point, The, 84
Tolkien, J. R. R., 158
TomTom NV, xv, 63
Torvalds, Linus, 139
transient advantage, 51–52
trends, recognizing, 48–49
TRIZ, 77–78, 149
Truscott, Bill, 143
Turkish towel, invention of, 109–10
Twitter, 65, 157

U

University of the People, 4
Unix software, 139

V

value, versus benefit, 12
value proposition, 112–13, 193
   developing, 114–15
values
   and behavior, 125–30
   maintaining and strengthening,
      131
Virgin Atlantic Airways Ltd., 79
vision
   versus challenge, 172
   communicating, 218–20
Vodafone, xv
Volkswagen, 87
Vouchercloud, 50, 61

W

Wang Laboratories, 139
WaterCan (WaterAid), 61
Watson, Thomas, 97
Watson Sr., Thomas J., 181
WD-40, 128–29, 150
Webb, Brendan, 133
WestJet Airlines Ltd., 79
white space, market, 49
Wilson, Harold, 4
wisdom of crowds, 24, 65, 225
Wonder, Stevie, 27
work hard/play hard culture, 124

X

Xerox Corporation, 85, 174

Y

Yellow Tail Shiraz, 41, 60, 73
YouTube, 84

Z

Zorbit nappies, 215

The Knowledge Center
www.asq.org/knowledge-center

Learn about quality. Apply it. Share it.

ASQ's online Knowledge Center is the place to:

- Stay on top of the latest in quality with Editor's Picks and Hot Topics.
- Search ASQ's collection of articles, books, tools, training, and more.
- Connect with ASQ staff for personalized help hunting down the knowledge you need, the networking opportunities that will keep your career and organization moving forward, and the publishing opportunities that are the best fit for you.

Use the Knowledge Center Search to quickly sort through hundreds of books, articles, and other software-related publications.

www.asq.org/knowledge-center

PUBLICATIONS   The Global Voice of Quality™

Ask a Librarian

Did you know?

- The ASQ Quality Information Center contains a wealth of knowledge and information available to ASQ members and non-members

- A librarian is available to answer research requests using ASQ's ever-expanding library of relevant, credible quality resources, including journals, conference proceedings, case studies and Quality Press publications

- ASQ members receive free internal information searches and reduced rates for article purchases

- You can also contact the Quality Information Center to request permission to reuse or reprint ASQ copyrighted material, including journal articles and book excerpts

- For more information or to submit a question, visit **http://asq.org/knowledge-center/ask-a-librarian-index**

Visit www.asq.org/qic for more information.

TRAINING   CERTIFICATION   CONFERENCES   MEMBERSHIP   **PUBLICATIONS**

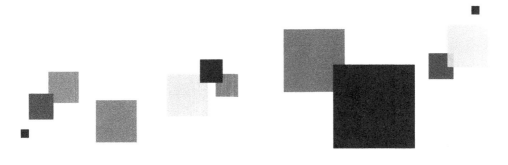

Belong to the Quality Community!

Established in 1946, ASQ is a global community of quality experts in all fields and industries. ASQ is dedicated to the promotion and advancement of quality tools, principles, and practices in the workplace and in the community.

The Society also serves as an advocate for quality. Its members have informed and advised the U.S. Congress, government agencies, state legislatures, and other groups and individuals worldwide on quality-related topics.

Vision

By making quality a global priority, an organizational imperative, and a personal ethic, ASQ becomes the community of choice for everyone who seeks quality technology, concepts, or tools to improve themselves and their world.

ASQ is...

- More than 90,000 individuals and 700 companies in more than 100 countries

- The world's largest organization dedicated to promoting quality

- A community of professionals striving to bring quality to their work and their lives

- The administrator of the Malcolm Baldrige National Quality Award

- A supporter of quality in all sectors including manufacturing, service, healthcare, government, and education

- YOU

Visit www.asq.org for more information.

TRAINING CERTIFICATION CONFERENCES MEMBERSHIP **PUBLICATIONS**

The Global Voice of Quality™

ASQ Membership

Research shows that people who join associations experience increased job satisfaction, earn more, and are generally happier*. ASQ membership can help you achieve this while providing the tools you need to be successful in your industry and to distinguish yourself from your competition. So why wouldn't you want to be a part of ASQ?

Networking

Have the opportunity to meet, communicate, and collaborate with your peers within the quality community through conferences and local ASQ section meetings, ASQ forums or divisions, ASQ Communities of Quality discussion boards, and more.

Professional Development

Access a wide variety of professional development tools such as books, training, and certifications at a discounted price. Also, ASQ certifications and the ASQ Career Center help enhance your quality knowledge and take your career to the next level.

Solutions

Find answers to all your quality problems, big and small, with ASQ's Knowledge Center, mentoring program, various e-newsletters, *Quality Progress* magazine, and industry-specific products.

Access to Information

Learn classic and current quality principles and theories in ASQ's Quality Information Center (QIC), *ASQ Weekly* e-newsletter, and product offerings.

Advocacy Programs

ASQ helps create a better community, government, and world through initiatives that include social responsibility, Washington advocacy, and Community Good Works.

Visit www.asq.org/membership for more information on ASQ membership.

*2008, The William E. Smith Institute for Association Research

TRAINING CERTIFICATION CONFERENCES **MEMBERSHIP** PUBLICATIONS